POLE AND TOWER LINES

McGraw-Hill Book Co., Inc.

PUBLISHERS OF BOOKS FOR

Electrical World ▽ Engineering News-Record
Power ▽ Engineering and Mining Journal-Press
Chemical and Metallurgical Engineering
Electric Railway Journal ▽ Coal Age
American Machinist ▽ Ingenieria Internacional
Electrical Merchandising ▽ BusTransportation
Journal of Electricity and Western Industry
Industrial Engineer

Pole and Tower Lines

for

Electric Power Transmission

By

R. D. Coombs

ISBN: 1929148429

WEXFORD COLLEGE PRESS

ΩΧΠ

THE MAPLE PRESS · YORK PA

PREFACE

The rapid growth of the electric light and power industry within the last decade has caused an enormous increase in both the number and length of transmission and distribution lines. From a comparatively unimportant detail—of little interest even to the owner—the power lines have developed into quite pretentious systems whose relation to other industries and to the public must be taken into consideration. The occasional transmission systems designed for 11,000 volts and consisting of wood pole lines with spans of 120 ft. have developed into 150,000-volt lines on steel towers with spans from 500 to 800 ft. long, while the light wooden poles supporting a few street lighting circuits have been superseded in some cases by very heavy trunk lines of many cables. As is to be expected in a growing industry, no set of standards has been universally adopted. Moreover, it is impossible to apply any one specification or standard of construction universally, unless such a standard has some elasticity and is interpreted and enforced intelligently. In any attempt at standardization, either for one operating company or in a national specification, it is necessary to consider carefully the cost and the operating problems involved in the adoption of any general mechanical requirements. An apparently harmless requirement may become a very serious matter if applied to all future construction.

The number, size and voltage of wires and the type of insulation to be used belong to the field of electrical engineering, while the subsequent determination of the best method of carrying the conductors across country is purely a question of civil engineering. The terms "electrical" and "civil" are used in their narrow sense, but the division is important as denoting the proper apportioning or departmentizing of the work. In this age of specialists there is no more reason for electrical engineers to determine structural details, than there is for structural engineers to decide on the proper type of insulator. If laymen or workers in other branches of the engineering profession have hesitated to assume authority in electrical matters, the same cannot be said of the average electrical expert in the field of pure construction. This, no doubt, is due to the fact that those engaged in coordinate branches of the profession have done little or nothing to solve the problems properly assignable to construction engineers.

That this is no imaginary charge is substantiated by the history of transmission line construction. Until recent years the rule-of-thumb practices of telephone companies—worthy results of the test of time as many of them are—were followed blindly by those in charge of electrical transmission. As a result there are many improperly constructed lines and erroneous ideas prevail regarding the facts and principles involved in their design. This condition should no longer be allowed to exist.

It is not the purpose of the writer to deal with purely electrical problems, such as the relation of voltage and size of the wires to the electrical characteristics of a line, or with such very specialized matters as the design of insulators. The problem is rather to develop a clearer preception of the application of the laws of mechanics to the case in hand.

The writer wishes to acknowledge, with thanks, the assistance rendered by Mr. W. L. Cadwallader in preparing the tables and computations, and by those in charge of various properties in furnishing illustrations and data relating thereto.

R. D. COOMBS.

NEW YORK, N. Y.
1916.

CONTENTS

CHAPTER VIII

CHAPTER IX

CHAPTER X

CHAPTER XI

CHAPTER XII

CHAPTER XIII

CHAPTER XIV

POLE AND TOWER LINES

CHAPTER I

GENERAL CONSTRUCTION

The choice of a type of construction for transmission lines, on private rights-of-way, may be said to depend on the cost of construction *versus* the cost of maintenance and of interruptions to service. This is also true, though to a less extent, of certain classes of railroad power lines. Where the lines of a transmission company are on public property, or cross other lines, or when high-voltage lines are used by railroads for electric-traction, the cost of failures of construction will be found to exceed the additional cost of good construction.

A decision as to the exact type of line, character of supports, and total wire capacity which will provide the highest ultimate economy, as well as excellence of service is perhaps impossible. A designer can only use good judgment in estimating future developments, and engineering skill on the immediate construction. The number of years that any particular line, perhaps any or all lines, will need to remain in the original location cannot be definitely foretold. If the line remains in place for a long period of years, will it have any considerable scrap value at the end of that period? If removed, or considerably altered, in a shorter period, the value remaining in the material is scarcely less difficult to determine. Apart from the economic needs of the owner 10 or 20 years hence, there is the question of the effect upon the existence of the line, of federal or municipal regulation at some earlier date.

In addition to these considerations, there must also be taken into account the probable future wire-carrying capacity and voltage, the life of the materials, and finally the relative ultimate economy of one or another type of construction as judged by the above conditions.

In order to obtain a better understanding of the various factors which affect, from a construction standpoint, a decision as to the type of line, the following details may be noted:

1

Carrying Capacity of Supports.—Experience with the lower voltages, at least, has shown that the capacity of the supports to carry additional wires has usually been underestimated. The extremely heavy distribution lines often seen on city streets are not, however, an example of a proper ultimate wire capacity, but rather of an unwise overloading. Exact maximum limits cannot be set, but the occasional failure of such overloaded lines

FIG. 1.—Overloaded wooden poles.

FIG. 2.—One-circuit wooden poles, 33,000 volts.

causes an expenditure which should be a weighty argument for the alternative of more and lighter lines. It is true that two light lines will cost more than one heavy line in so far as the actual construction cost is concerned, but the increased maintenance of the latter and its greater liability to extensive or prolonged interruption to service, if difficult to foretell in terms of money, are not less costly on that account. Many otherwise competent engineers seem to ignore the very real cost in time, money, and reputation "saved" by a reduction in the initial construction account.

The ordinary wood-pole distribution lines are subject to more alteration than transmission lines, and are affected by conditions not generally applicable to transmission lines. In transmission-line construction, the question is whether to provide for one or two, two or four, or four or more circuits upon a single line of supports, and in the higher voltages whether to carry one or two circuits.

The financial ability or disability of the owner may settle the question of the number of circuits without reference to any further considerations whatever. The best may be the cheapest, but it may also be the impossible. Under such circumstances, the engineer can build only as well as is practicable with the funds available. Lines built under such restrictions are usually in less populous territory, and the penalties of accident are less severe than in the neighborhood of cities. The natural and entirely proper practice will be to provide relatively low supports, long spans, light structures, and very little future capacity. It may be noted, however, that proper side clearances, spacing, insulation, and guying, add but little to the cost compared to the gain in the efficiency of the line.

In what may be called the better grades of construction, the number of circuits upon a single support depends very largely upon the location of the line and its present or future voltage. There is a quite important difference between lines upon private rights-of-way and those upon highways or private property. It is not unreasonable to assume that an operating company may utilize any combination of heavily loaded or duplicated lines it chooses upon its own property, though that property will eventually have to be fenced and patrolled. On the other hand, the occupation of highways, etc., by overloaded or perhaps any lines, may become subject to criticism. Again the holding of separated parallel, or looped, lines has real advantages from the commercial and service standpoint.

Dividing the total capacity for wires between two or more pole lines cannot be justified as a measure of reducing the cost. Such duplicate lines will always cost more than one heavy line, but the subdivision may be justified on other grounds. In case one line is given a different route, the maximum security from simultaneous interruption to service is obtained and additional busi-

FIG. 3.—Two-circuit wooden poles, 33,000 volts.

ness may be secured, but the cost will be more than that of two lines on one right-of-way, and nearly double that of one line of the total capacity.

The number of circuits that may be carried by a single support depends on the electrical characteristics, the continuity of service desired and to some extent on the length of span and the character of the supports. Thus the same weight of conductor divided between two circuits affords considerable protection against interruptions to service, with little added cost except for higher poles, larger or more numerous arms, more insulators, etc. A single circuit, however, can be arranged with more space between conductors and may be used more economically in long spans.

Life of Materials.—In general it may be assumed that a line should have the longest life consistent with a reasonable first cost, and that after selecting a type of construction and factors of

Fig. 4.—Two one-circuit lines, 55,000 volts.

safety proper for the location under consideration, the matter of restrictive regulation must be left to the future.

In connection with the probable life of materials the statement is frequently made that the life of steel is indefinite. This is literally correct though not exactly as intended. The life of properly protected steel is indefinitely long, while that of steel exposed to the weather and unprotected is indefinitely short. Many galvanized wind mills which are 20 years old are still in good condition, and some painted steel structures are much older. On the other hand, there are both galvanized and painted structures whose probable life will be less than 20 years. The probable average life of unprotected timber poles is about 10 years, while that of poles which have been properly treated with preservatives is from 16 to 20 years, depending on the treatment. As reinforced-concrete poles are a more recent development there has not been the same opportunity to obtain fair averages, but

Fig. 5.—Two one-circuit steel towers, 60,000 volts.

from observing installations 10 years old it appears that their average life is comparable with that of steel.

Clearances.—Until recent years the only general clearance specified by engineers, or required by law, was that power wires should be maintained at a height of 25 ft. above highways or above the top of rail over railroad tracks. In some cases a clear height of 22 ft. was permitted. Such clearances, like most of the early requirements in transmission construction, were founded on and often a direct copy of existing telephone and telegraph practice.

The overhead clearance necessary to permit the trolley pole on a large modern car to swing into its upright position, in case it

Fig. 6.—Two-circuit steel towers, 110,000 volts.

accidently leaves the trolley wire, would perhaps indicate the minimum limit. While overhead systems for alternating-current operation of railroads or interurban lines are, as yet, but few in number—the largest and best known installation being that on the New York Division of the New York, New Haven & Hartford Railroad—such construction should receive consideration in establishing a standard overhead clearance above railroad tracks. In such cases the railroad company would probably desire to have their transmission circuits, if on a separate pole line, carried above all crossing lines, except possibly those of very high voltage systems. However, railroad trolley-contact wires cannot be elevated above other crossing wires. The minimum height of such overhead contact wires is approximately 23 ft. at the center of a span and 28 ft. at the supports.

Linemen working on foreign wires, whether on joint-pole lines or not, should be protected from contact with power wires by providing a reasonable space between the two lines or sets of wires. In this connection mention may be made of one reason for requiring telephone and telegraph wires to be placed below power wires either at crossings or when located on joint poles, *i.e.*, to prevent the harmless wires dropping into contact with the power wires, either during the process of stringing wire or due to the more frequent mechanical failure of the smaller telephone wires.

It should be remembered that the stresses on poles increase directly with their height and, in fact, more rapidly than in a direct ratio when there is no adjoining protection from the wind. Omitting considerations of accidental contact and malicious injury, the lower line is usually the safer line. Assuming that the power or lighting circuit is, as it should be, the higher circuit, or in the case of several lines of different voltages that they are arranged in order with the highest voltage circuit uppermost, there should remain below the power wires a zone for harmless wires. This arrangement is especially necessary over highways or railroads where such inferior lines exist or may be reasonably expected in the near future.

The proper separation of conductors to prevent their swinging into contact has never been definitely determined. It has been argued that long spans swing synchronously, and that experience has shown that they may be safely spaced much closer together than the distance required to provide for the maximum displacement. On the other hand, short spans with relatively greater separation have been brought into contact by what appears to be a purely electrical cause.

It is doubtful whether engineers and executives realize the extent to which undesirable construction is installed, due to inadequate efforts to obtain the necessary concessions on the part of outside interests. Isolated trees, of perhaps no particular value, have compelled the use of high poles or the unnecessary grading up of pole lines. Without wishing to appear an advocate of some of the common methods of "tree trimming," the writer believes that one large scraggly tree, more or less decayed, remaining along a curb line where the other trees are of recent

Fig. 7.—Steel pole line, 11,000 volts, with space for lower voltages.

regular growth, should not be allowed to interfere with the proper location of all the wires in that street.

Everywhere throughout the country, there are towns in which the telephone and electric service lines occupy all sorts of zones, gradually getting higher and higher, until the latest line is driven to pole heights extremely difficult to obtain. It will also be found that many telephone lines have overbuilt the present

power wires and occupy the zone which should be used by future
power wires of higher voltage. Telephone or telegraph wires
should be placed underneath power wires, as it is impracticable
to give the former a sufficient factor of safety to prevent mechanical failure.

There is no established relationship between the span and
sag, and the separation between conductors. Two general
methods have been advanced for determining the separation between conductors: (1) depending on sag and span, and (2) on
sag only. The first is incorporated in the Pennsylvania Railroad
wire-crossing specifications (1914) which prescribe a separation

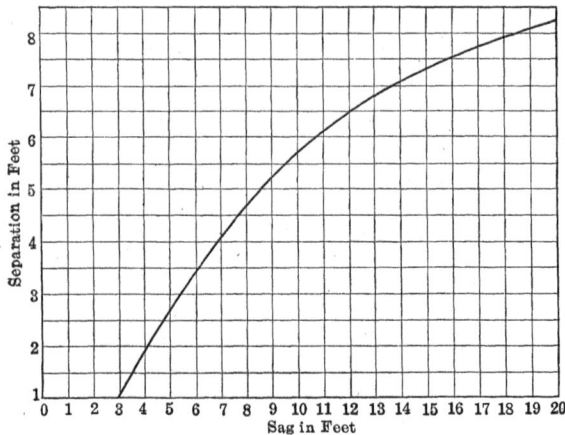

Fig. 7 *a.* —Conductor separations.

of 1 in. for each 20 ft. of span plus 1 in. for each foot of sag or
fraction thereof. The second method, proposed by the writer
in a paper before the National Electric Light Association con-
vention in 1913, is shown in Fig. 7 *a.*

In both methods mentioned, the minimum separation may not
be less than that given for shorter span lengths in paragraph 3
of the specifications on page 259. The use of suspension insulators involves an additional separation of one and one quarter
times the length of the string of suspension insulators.

Tree Trimming.—In order to provide and maintain adequate
separation between conductors and adjoining timber, a rather
indefinite amount of tree cutting and trimming must be done.
The actual amount of such work will vary between wide limits

for different lines even in the same locality. For lines located on streets the problem is first to select the most accessible and least-shaded street, and second, to adjust the height of the poles and the amount and character of trimming so as to obtain the maximum protection with the minimum of offense. In cross-country lines, however, particularly those on private right-of-way, the problem is somewhat different. In this case, there is usually some freedom of movement through which the line may be so located as to avoid too close proximity while retaining the benefit of distant shelter. This latter feature seems to have been generally disregarded, and yet, provided sufficient separation is maintained to prevent falling contacts, the presence of timber land to windward is in the writer's opinion a considerable asset in the strength of ordinary lines.

Cutting down trees, while generally deplorable, cannot always be avoided, so its justification must necessarily depend on the quality of the interfering tree and the importance and position of the line. Some trees will so outrank the ordinary power line, both in their real and in their popular value, that a change in route may have to be considered. Other trees are past their prime and have merely a sentimental value to a limited number of persons. In some cases permission for indiscriminate cutting will be freely given. Either in cutting or in trimming, a broad-minded liberal policy on the part of the power company, coupled with considerable tact, will ultimately justify itself.

Unpruned trees with long scraggly limbs, instead of being injured, will generally be improved by proper trimming. Dead or dying branches are of no benefit to trees, whereas they are a serious menace to the power company; therefore they should be removed in the immediate neighborhood of the line.

The methods in vogue in trimming trees are greatly in need of improvement. Aside from the serious loss in popularity and prestige, it is nothing short of criminal waste to unnecessarily injure grown timber. This country once possessed enormous forests, yet the present timber lands are only a pitiful remnant and cultivation is almost unknown. In pole-timber land it is no exaggeration to claim that every tree unnecessarily cut down or killed adds its mite to the future maintenance expenditures of the local power company.

By exercising some care it is possible to trim so that killed trees or non-permanent clearances should be rare. The season of the year in which trimming is done has a marked influence on the successful healing of the cuts.   In general it is best to trim any tree during the dormant season.

In removing large limbs they should be first undercut to prevent slivering and then sawed through close to the trunk.   The stump should then be cut off *flush* with the trunk, leaving a *smooth* surface, which may be painted when it has dried.   If this is not done, the stump will decay and the rot will spread to the trunk.   Small limbs may be cut immediately beyond any forks, or close to the trunk; upright limbs should be cut on a slant and the surface should be finished smooth and then painted.

When an entire tree is to be cut down, the stump should be short and be given a smooth ridge or roof similar to that on a pole top.   Second-growth trees will then usually sprout from the stumps and be available for lumber in the future.

In protecting or patching rotten cavities, the dead wood should be cut away to form a pocket with undercut edges as in dental work, and the cavity should then be painted and left as an open cavity or be filled with cement mortar or asphalt.

On private right-of-way all trees, brushwood, sage brush, etc., should be cut down and cleared away.   The amount of cutting on neighboring property will vary greatly, but should in all cases include the trimming of nearby dead branches and such trees as by their unusual position create a particular hazard.

The attachment of guys, without tree blocks or shields, will frequently kill a part or all of a tree, lessen its value as a stub, and also render it a menace to the line.

Trees grow by the addition of wood fiber to the outside of existing limbs, and the new growth is fed by a descending liquid in a thin layer, called the cambium, immediately inside the bark.

If, therefore, the continuity of the cambium layer is broken the new wood is starved just below the break.   If the cambium of all or a large proportion of the circumference is cut through or is prevented from growing with the tree, the entire tree starves.

Nearly all trees are dormant in winter and may then be cut without excessive bleeding and without subjecting the wounds to attack by insects.

Right-of-way.—The simplest form of right-of-way is that of pole or tower rights, whether obtained along streets by franchise from the municipality or by lease or purchase from private owners.   Although this form is by far the most common, it seems probable that there will be a marked increase in the number of private rights-of-way for both pole and tower lines. Private rights-of-way, while naturally more expensive in original cost, permit the use of the most economical types of construction and provide insurance against restrictive regulation or an excessive cost for increased facilities. The abnormal expenditures involved in hurried construction and the excessive payments often required to complete a right-of-way—a species of blackmail—are perhaps not fully realized. Such expenses would be greatly reduced in the case of a private right-of-way, particularly for subsequent lines installed thereon. It might be argued that excessive payments would be demanded for a continuous right-of-way, as is usually the case in railroad construction. While such unit prices are unquestionably excessive for the land as such, they may not be excessive for an electric right-of-way—at least this has been the case with railroads. Heretofore, private rights-of-way have been purchased chiefly where land was very cheap and when an important line on wide-base towers was to be constructed. It is probable, however, that equally effective reasons may be advanced for private right-of-way in more settled communities, on which to build a series of pole lines.

There seems to be no general standard or set of rules by which the width of a right-of-way may be determined. The two factors which appear to have had the greatest influence are the height of adjoining timber and the probable ultimate number of pole or tower lines. In addition to these, there are several other conditions which should affect the width: the total desired security of the lines; the character of the country traversed; and the character of the construction.

Where the nature of the ground permits, some consideration should be given to patrol or transportation facilities between the lines of supports. Apart from any question of cultivation or possible railway facilities, the remaining conditions are all involved in the general one of security from interruptions. Interruptions may originate either within or without the limits of the right-of-way. Those from within are generally due to mechanical or electrical failure and are best minimized by separating the lines. Those from without—which usually exceed the former—include falling trees, limbs, straw or other objects blown by the wind, fires, malicious mischief, etc. They are minimized by moving the supports in from the side lines. The two types of interruption are therefore prevented by opposite action, but since the latter set is more important it is advisable to give them more weight in the location of the poles or towers. It has sometimes been stated that the distance from the side lines should equal or exceed the height of the tallest neighboring trees. Literally applied this rule would require a variable width, and in some localities excessive widths. While some degree of consideration may be properly given to the average height of timber, it must not be forgotten that the effective range of wind-blown branches is too great to permit absolute protection. It is advisable to cut down or trim the taller trees and to remove dead branches, since storms would presumably blow these onto the line before the smaller and live timber was affected.

The character of the construction will affect the separation of the supports and their distance from the property lines because the wind-blown sag of the wires must be given proper clearances, either from each other or from the side lines. Long-span construction will, therefore, require greater side clearances than short-span construction. Steel poles or narrow-base towers permit the closest spacing of the lines, both on account of their narrow spread at the ground and because they are usually employed with shorter spans and smaller sags. If the supports are staggered, *i.e.*, the poles in one line opposite the center of spans of the adjoining line, less clearance is needed to prevent swinging contacts. The general security desired will affect the width of the right-of-way and the location of the lines thereon. One line in the middle of a wide right-of-way has the maximum possible security. In wild, treeless country two lines near the edges of the property are more immune against interruptions than with a smaller separation. One tall and one low line are more immune than two tall lines because the lower line can rarely affect the taller. For all other conditions, the lines should be located so as to permit the greatest freedom of future construction, a reasonable separation between lines, and a maximum clearance from the side lines.

In order to facilitate the study of right-of-way clearances, three types of installation are shown. Fig. 8 represents a two-circuit steel pole or narrow-base tower located in the middle of a private right-of-way; Fig. 9 shows two one-circuit poles, while Fig. 10 shows two two-circuit wide-base towers. It is assumed that suspension insulators have been employed in each case.

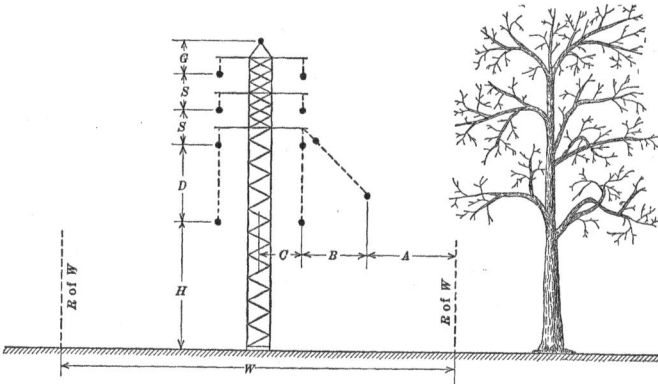

Fig. 8.—Right-of-way—one two-circuit pole line.

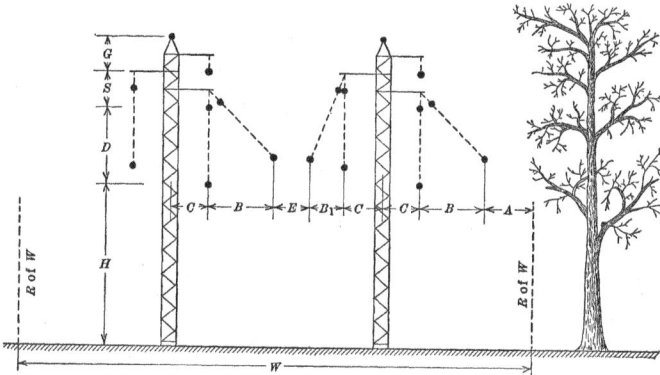

Fig. 9.—Right-of-way—two one-circuit pole lines.

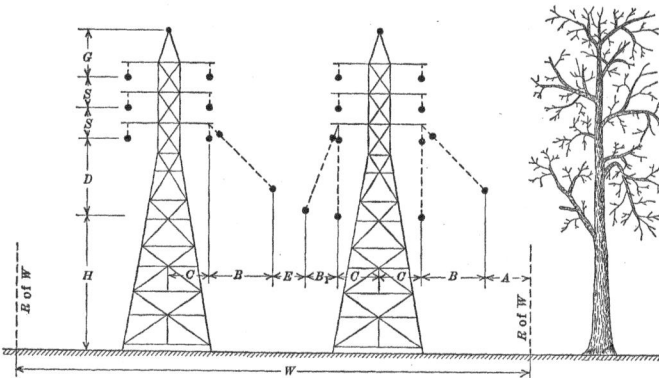

Fig. 10.—Right-of-way—two two-circuit tower lines.

To modify the diagrams for pin insulators it would only be necessary to adopt a slightly smaller value for the wind-blown deflection B and for the tower clearance C.   The windward-deflected
sag B_1 in Figs. 9 and 10 is shown as one-half that of the leeward-
deflected sag B.   This assumption is, of course, arbitrary and
would vary for different sizes of wires, spans and sags.

In Table 1 are given the probable ranges of the various clearances, the summation of which determines the width of the right-
of-way.   Also shown therein is the probable range of the width
W with three assumed minimum side clearances A, of 6 ft., 15 ft.,
and 25 ft.

The presence of tall trees touching the right-of-way line is undesirable, but it cannot always be avoided.   The diagrams represent the limiting condition.

TABLE 1.—APPROXIMATE RANGE OF RIGHT-OF-WAY CLEARANCES (IN FEET)

	One double-cir-cuit pole line	Two single-cir-cuit pole lines	Two double-cir-cuit tower lines
Span....................	300 to 500	300 to 500	500 to 800
H......................	20	20	20
B......................	3 to 12	3 to 12	10 to 25 .
B_1......................	....	1 to 6	6 to 12
C......................	3 to 6	3 to 5	4 to 8
E......................	....	3 to 5	3 to 6
If A = 6 ft., W =.......	25 to 50	35 to 80	65 to 135
If A = 15 ft., W =.......	40 to 65	55 to 95	85 to 155
If A = 25 ft., W =.......	60 to 85	75 to 120	105 to 175

Factor of Safety.—As in all construction work, by far the most
important determination is the mechanical factor of safety.
Indeed, the factor of safety will not only narrow the selection of
the details of construction, but will practically determine the
general type of line to be used.   In past practice, however, the
factors of safety have frequently been the last values to be definitely determined, whereas they should be the basis for computation.   In other words, the method in general use is a cut-and-try
method, in which several designs having in reality different factors are first worked out, and the selection too often left to one
individual's judgment or to the suggestion of a salesman.   It is
extremely doubtful whether "competitive" designs received by

most purchasers have been really comparable, in so far as their true mechanical factors of safety were concerned. In addition to the vagaries of competitive bidding, it may be claimed with considerable justice that the designs ordinarily made by a purchaser are not truly comparative. For an accurate survey of the conditions influencing the selection of the proper factors of safety for the various members involved, it is essential that consideration be given to the desired length of service of the line, as well as to the characteristics of the members and materials involved in the construction. Included in the term "length of service" are many indefinite quantities which must be determined by judgment. Changes in the capacity of the line, a possible increase in voltage, or the entire elimination of the line from an operating standpoint may perhaps be assumed with some degree of accuracy, but the probable life of the materials of construction and the possibility of restrictive public regulation are extremely difficult to determine, either for a given line or for future developments as a whole.

The factor of safety, or as it is sometimes termed, "factor of ignorance," is a much-abused and generally misunderstood expression. In reality it is a combination of the allowances for error, and consists of the summation of the individual allowances or elements of the factor, the term safety being somewhat misleading. The amount of the total factor depends, or should depend, on the accuracy with which the conditions of service and the characteristics of the members and material can be foretold. If the possible variation of all the individual elements except one is known, then the allowances for the known elements entirely eliminate them in any further consideration of "safety," and a further increase in the total factor causes a disproportionate increase in the one unknown. For example, the factor of safety for wires may be sub-divided into the following elements:

(*a*) Increased loading.
(*b*) Uncertain strength of the material.
(*c*) Injuries during erection.
(*d*) Errors in erection (improper sag).
(*e*) Deterioration in the material.

While the theoretical values of these elements should vary for each installation, the actual general case may be stated as approximately:

2

$$a = 0.30$$
$$\cdot b = 0.20$$
$$c = 0.10$$
$$d = 0.30$$
$$e = 0.10$$

$$\text{Total} = 1.00$$
$$\text{Breaking strength} = 1.00$$

$$\text{Total factor of safety} = 2.00$$

In other words, an analysis of the commonly used factor of 2.0 shows that it permits the loading to be underestimated 30 per cent., and the actual breaking strength of the wire 20 per cent., and allows a decrease of 10 per cent. from small injuries, an increase in stress of 30 per cent., due to improper stringing, and an ultimate deterioration of 10 per cent., before the span is theoretically at the point of failure.

Omitting for the moment any consideration of the elastic limit, and the fact that stresses in excess thereof will necessitate pulling up slack (with or without other undesirable results), it is evident that any further increase in the factor of safety will greatly increase the allowance for the most uncertain element. On the other hand, if the strength of the wires assumed in the design corresponds closely with the material as purchased, and the wires are strung with care and with sags having a close approximation to those in the design, it is apparent that the spans will safely withstand a very considerable increase in the assumed loading.

A little consideration of the probability of exceeding the above elements in a line designed and erected with reasonable care may explain the excellent record of the wires in existing lines. It is more than probable that in many instances inaccurate wire stringing has entirely changed the actual strength of the wires in relation to external loads, and as indicated by the allowance of 0.30 in the above analysis, this is the most uncertain condition in the average installation.

The above analysis is, in the writer's opinion, a fairly accurate statement of the average actual condition, but does not give the correct values for the elements of the factor of safety of wire spans designed and constructed under competent supervision.

Spans.—Theoretically, since the cost of the material between supports, *i.e.*, the wire, is constant, except for the slight increase in length due to the sag, the supports should be spaced far apart. With long spans the number of insulators is reduced, together with the probability of interruptions originating at the supports. However, other considerations usually prevent the adoption of the theoretically economic span length. The conductors must be spaced so as to provide sufficient clearance between adjoining wires and between the wires and the pole. Therefore with an increase in span length, with its consequent increase in sag, it becomes necessary to spread the wires further apart, thus lengthening the crossarms and increasing their cost. With comparatively few wires in the line, it is possible to arrange them so that long spans can be used without excessively long or heavy crossarms, but on heavy lines carrying many wires, this is not practicable without unduly increasing the height of the poles.

There is quite a difference between the meaning of "average span" and "standard span," the former being the final result and the latter the original design which provides a theoretical clearance above the ground in flat country. Therefore, unless the towers can occupy hill tops, or an extra clearance is allowed in the design, it frequently happens that intervening elevations, or the loss of clearance on hillsides, materially decrease the actual span length.

If lines are located on highways long spans are not always practicable as the length of crossarms may have to be restricted, or the wooden poles available may be incapable of withstanding the load due to long-span construction.

The matter is further complicated by the mechanical limitations of standard, or stock, crossarms, pins, and insulators.

On steep hills the spans must be decreased, or the supports lengthened, to maintain the overhead clearance.

The size of the conductors has more effect on the proper or possible length of span than any other condition, since the large sags required for small wires in long spans would necessitate excessive wire spacings and pole heights.

No exact economic span length has been determined either for one type of support or for one section of country. In fact, it is extremely probable that for any particular line there will be two designs of nearly the same estimated cost, and that the possible error in estimating the field work will far exceed any difference between the estimates of material.

In wood-pole construction, the use of long spans with high

poles is subject to a serious error in estimating the probable replacement cost. In view of the recent prices and scarcity of long poles, it may be possible that such lines cannot be rebuilt in timber at any reasonable cost. The wood-pole transmission line is an entirely proper type of construction in many cases, and it is also true that for one or two circuits the spans could often be lengthened with advantage, but such lines should be protected from decay and a high replacement cost used in estimating.

While the economic design in traversing hilly country is undoubtedly to cross ravines and small valleys by means of long spans, it is possible that this practice may be injudicious unless ample clearance is provided between the wires. There is little exact knowledge of the dependence to be placed on the parallelism of swinging wires, particularly if their horizontal spacing is only 5 to 10 ft. and the sag 15 to 30 ft. Besides the accidental contact of wires in the same horizontal plane, there have been instances of the lower wires being lifted by the wind into contact with those above.

In case it will be necessary to pay rent for pole-rights on foreign property, it may prove economical to use long-span construction as the rent saved might more than pay interest on the increased cost of the supports.

In ordinary country, the economic span is probably between 400 and 500 ft. for narrow-base supports, and between 600 and 800 ft. for wide-base towers.

Supports.—The poles or towers used up to the present time have been of wood, steel and reinforced concrete, and they have been used in the order given, both as to numbers and priority of installation. Wood poles, still the most common form of support particularly for low-voltage lines, have several objectionable features in that they deteriorate rather rapidly, do not resist fire and their cost is increasing. Under certain conditions, however, wood poles are still economically sound construction even for high-voltage lines, although the time is not far distant when they will no longer be employed for first-class installations.

In changing from wood to metal, however, we may profitably pause to consider some of the characteristics of the structure which has rendered possible our progress in line construction. In theory as well as in fact the wooden pole is a precedent for the metal structure, and a too violent divergence from some of its good features may result in structures not relatively so excellent as the wood they replace. A well-selected timber pole is very nearly of the ideal outline, due to the fact that the stresses imposed upon it in its original life were almost identical in nature with those encountered in pole-line service. It should not be forgotten that a wood pole has equal strength in all directions, both with and across the line, and a comparatively large strength in torsion. These qualities tend to minimize the effect of acci-

FIG. 11.—Bending test, Coombs concrete pole.

FIG. 12.—Tower of the rigid or windmill type.

dental loads or loads other than those assumed in design. Again, wood poles have considerable elasticity but not complete flexibility, a characteristic which enables them to deflect enough to equalize most unbalanced loadings while opposing a very considerable restraining force against the spread of failures along the line. This semi-flexible quality of wood poles, which is also obtainable in steel or reinforced concrete, is probably of much greater advantage than is generally realized. Another advantage of wood poles is that they are not easily injured in handling, and may be installed by men of ordinary intelligence and training.

Moreover, there are no long thin sections which may be bent and rendered useless and no flimsy connections in the make-up of a wooden pole. That the above good qualities have been largely instrumental in securing the excellent record of wooden poles in line work cannot be doubted by the analyst, and their lesson is well worth attention.

The more permanent types of support may be divided into the rigid wide-base steel tower, the semi-flexible pole (either of steel or reinforced concrete), and the flexible steel frame. Apart from their relative cost for any given line, there is to be considered the ultimate adaptability of each type. This adaptability will involve the questions of protective coating, rights-of-way, freedom from serious interruptions to service, and finally, and to the writer's mind of considerable importance, the relative prominence given the installation.

In the progress of a rapidly growing industry there is always a tendency to apply methods of work to sections of the country to which they are less adapted than the locality of their previous successful use. The adoption of the so-called wind-mill tower may not be judicious in the densely populated districts of the East where climatic conditions are severe. In such regions, there are two considerations other than cost, *i.e.*, the undue prominence of the line and the great importance of failures. The wind-mill tower is rather conspicuous, but it provides a type of support with which failures are practically confined to one span. Flexible-frame supports are not so noticeable, but they have little or no strength in the direction of the line and will therefore presumably be susceptible to a more severe type of failure. The semi-flexible pole or tower, occupying a position about midway between the two types of structures mentioned, has at least a theoretical advantage over either. While some flexibility is useful in a narrow-base structure, to permit "pull back" by adjoining span wires, the amount of deflection need not be excessive. In actual service heretofore this movement cannot have been very great, for the reason that the commonly used attachments have not sufficient strength to transmit greatly unbalanced wire tensions. The desideratum is perhaps a certain elasticity rather than extreme flexibility. In fact, moderate bending or semi-flexibility is obtainable even in reinforced concrete.

In sparsely settled country or where the right-of-way is for any reason not accessible or not subject to cultivation, the spread of tower bases is unimportant. If more than one high-voltage line is to be placed upon a private right-of-way, the separation of the lines will usually depend upon factors other than the spread of the bases. When land is valuable wide-base towers may be impracticable. · For instance, there will probably be many power lines placed upon interurban railway rights-of-way. Such development is natural and necessary, but railroad rights-of-way do not provide space for wide-base construction.

FIG. 13.—Flexible A-frame.

FIG. 14.—Semi-flexible pole.

Wide-base towers and semi-flexible poles should, when properly designed, provide the maximum security against interruptions to service caused by insulator or wire failure. The greater strengths attainable in such structures allow the use of longer spans, with a consequent reduction in the number of insulators and of the probability of insulator failure. In case of wire failure, whether due primarily to insulator failure or not, the spread of such failure along the line is arrested before it has influenced more than a span or two. The remaining factor of adaptability, *i.e.*, the relative prominence of the various structures in the landscape, may prove of considerable subsequent importance. The writer does not mean to imply that a transmission line should be made

FIG. 15.—River-crossing tower.

decorative, but rather that it be made inconspicuous. Even in regard to decorative effect, it is not absolutely necessary that it be an ungainly blot upon the landscape. Some attention to pleasing outlines is not amiss, for it is a well-known fact that a gracefully designed structure is usually economical. In addition to the question of appearance, if lines situated in settled communities are to remain undisturbed for any considerable period of time, they will have to be either unobjectionable in performance or invisible.

The design of steel or reinforced-concrete poles and towers is fortunately becoming less hampered by demands for excessive cheapness, and the regulations current in other structural work are no longer entirely disregarded. The wisdom of this should be apparent when it is considered that a few hundreds or thousands of dollars "saved" on the line construction may jeopardize

Fig. 16.—River-crossing towers.

the efficiency of an investment of millions of dollars. It is true that thus far existing construction has given fairly satisfactory service, but it is equally true that the more extended use of faulty designs would eventually bring disrepute upon the industry, and through failures invite the enforcement of severe regulations by various authorities. In any type of support the importance of eliminating long unsupported members and of providing a firm rigid base is now becoming more generally recognized.

Location Plan.—After the general location of a line has been determined from a study of maps and inspection of the ground, the prompt completion of the location plan is essential. The rapidity of the compilation of this plan will depend on whether the line is to occupy, either entirely or in part, a strip of private right-of-way, highways, foreign rights-of-way, or pole-rights. In most cases, quite accurate preliminary data as to the plan view can be obtained from the right-of-way plans of properties such as steam or electric railroads, canals, highways, etc. When a private right-of-way has not been entirely secured, some changes in alignment may be expected, so the location plan is to that extent preliminary. Except for the desirability of having a correct permanent record, there is no particular need of determining by accurate survey the exact distances between widely separated points.

After the plan has been brought to a semi-final stage, the profile should be drawn upon the same sheet. In doing so, considerable future annoyance may be avoided by drawing a true profile and breaking the view at the corners, so that corresponding points will occupy their correct relative position in plan and profile, and both the center line and datum line will remain parallel to the bottom of the drawing. In some cases the distances in the profile have been measured along the inclined surface of the ground and then plotted horizontally, which results in a false profile and compels constant reference to plan and profile to identify a given point. It does not make a particle of difference in the excellence of a given section when completed whether the distance between the corners is 4000 ft. or 4050 ft. It is important, however, that the distance from small steep hills, etc., to one end of the section be accurately measured, so that the poles may be properly located to give the required clearance over the obstructions.

The tentative, or paper, location of the supports can now be made on the drawing and scrutinized in the field by walking over the line. It is assumed, of course, that in the preliminary location reasonable care was taken to avoid natural or artificial obstructions including side hills, swamps, flood lands, or undue interference with other lines, and the use of private property.

Almost invariably minor changes will have to be made to fit the paper location to the ground, and local surveys can be made to plot cross-profiles at side hills, crossings, encroachments, etc. Some supports will have fixed locations, *i.e.*, at the corners in the line, etc., so that the location of supports is reduced to distributing the poles or towers between the fixed points—a series of short

locations. In a line having many changes of direction and elevation, the previously assumed standard span length may seldom
be used. The problem is then one of ascertaining the economic
or desirable span length, not for a line in general, but for a given
series of short sections having fixed ends and various intervening
hills or other obstructions.

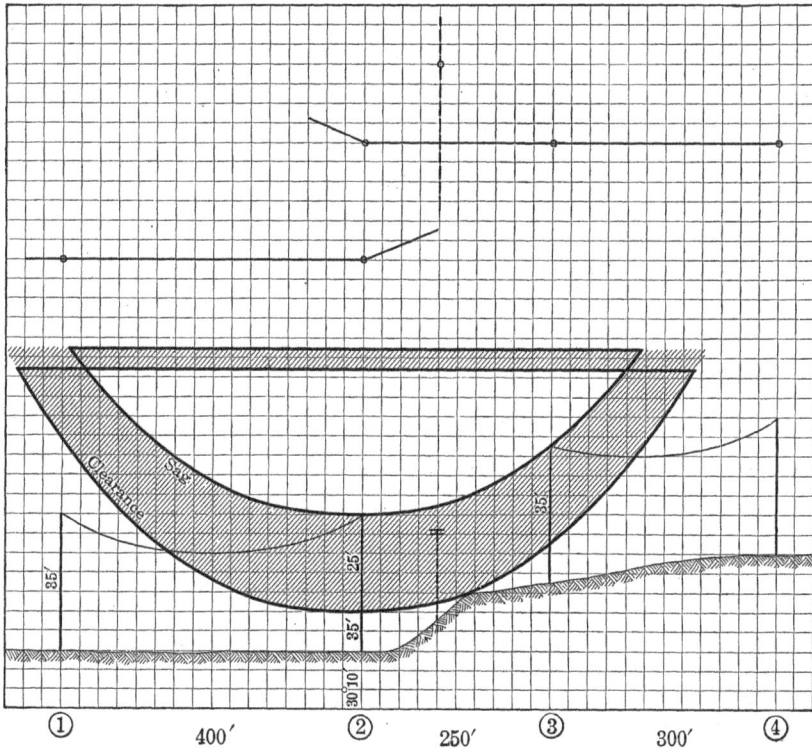

Fig. 17.

In drawing the plan and profile, it will usually be found convenient to use a horizontal scale of 200 ft. to the inch, and a vertical
scale of 20 ft. to the inch. All obstructions, highways, crossings,
etc., should be located in plan, and shown to scale in the profile.

A sag and clearance templet should be made of tracing cloth,
celluloid or even of thick paper, though the last is less convenient as it is not transparent. Such a templet is shown in
Fig. 17. The curve of sag is formed by plotting the maximum

sags of the given wire for various span lengths and usually for the condition of high temperature and no ice or wind load.

Parallel to the sag curve, and at the distance of the overhead clearance below it, is the clearance curve.

The templet should be extended to include span lengths well beyond the maximum span anticipated, in order that it may be used on hillsides.

The method of applying the templet is shown in Fig. 17, in which it should be noted that the base of the templet has been kept horizontal and the templet itself shifted until the sag curve coincided with the points of conductor attachment on two supports, without causing the clearance curve to intersect the ground line.

In case the standard height support in the assumed location will not permit this, the span must be decreased or the support be lengthened.

CHAPTER II

LOADING

No detail of line construction has been the subject of such inaccurate assumptions and misstatement of facts as the conditions of loading which actual existing supports should or would withstand. Medium-voltage lines, located in regions known to be subject to heavy sleet, have been described—and apparently designed—on the assumption that no sleet load would occur, and that 15 lb. wind pressure on the wires and 30 lb. on the supports were proper assumptions. It has been stated that, in some instances, provision has been made for the supports to safely withstand broken-wire loadings which have varied from one wire to one-half or two-thirds the number of wires. Of this entire set of assumptions, that of one broken wire combined with a proper wind loading is often a more accurate statement of the actual strength of the existing structures. Fifteen pounds pressure on a No. 0 bare wire is only 0.47 lb. per linear foot, while 8.0 lb. on $\frac{1}{2}$-in. thickness of sleet on the same wire is 0.91 lb. per linear foot. The pressure of 30 lb. per square foot, on the tower corresponds to a wind velocity of 112 miles per hour, which is so excessive as to provide a little extra strength in so far as that condition is concerned. The maximum tension in the wires would probably be about 2400 lb. per wire, and the usual pin insulators will not safely withstand such stresses. Furthermore, very few ties or clamps will dead-end a wire under such tension. Again, the crossarms used in certain of the lines under discussion would neither carry such unbalanced loads nor prevent a torsion which at less load would permit an insulator to incline and the wire to pull free. It is, therefore, evident that the assumed broken-wire stress could not be transmitted to the support.

The writer believes that no consideration need be given accidental loads caused by falling objects such as trees, etc., and that a single ice and wind load will apply very satisfactorily in nearly every part of this country. It is true that in certain localities either a smaller or a larger loading may be justifiable,

29

and that some installations may warrant greater security than others, but these are questions for engineering judgment and should not influence general construction.

Fig. 18.—Snow-ice loads.

Sleet.—As it is hardly practicable to attempt the consideration of accidental loads which can be caused by falling objects, the only external loads to be considered are the ice and wind loads

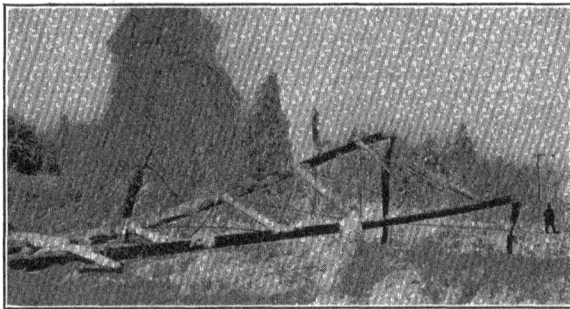

Fig. 19.

on the wires and their supports. Severe loads of this nature are rare, and those producing very excessive stresses may be regarded as being in the category with tornadoes and similar visitations which are beyond the limits of design. It has been shown by the records of the telephone companies, and is now more generally understood, that sleet loads may be encountered throughout nearly the entire United States, with the possible exception of certain restricted localities in the South and West.

The maximum amount of sleet undoubtedly varies, but the effective variation of the combined wind and ice load is much less than is generally believed. Further, and neglecting the occasional extremely heavy deposits, it seems probable that a maximum thickness of 1 in. may be encountered. Experience has shown, contrary to the earlier assumption of many engineers, that sleet deposits will occur on wires carrying voltages up to 60,000 and possibly much higher.

The heavier deposits are often of snow-ice and of less weight than clear ice, besides being more subject to removal by the sun and wind. Ice deposits, on the other hand, often remain intact even under a bright winter sun and a rising wind.

It is extremely improbable that every span in any given line will ever be subjected to the simultaneous action of the maximum sleet and wind loads. The maximum sleet load, provided for in the design, is in itself a rare occurrence for any given span, perhaps happening once in ten years. Moreover, it is assumed

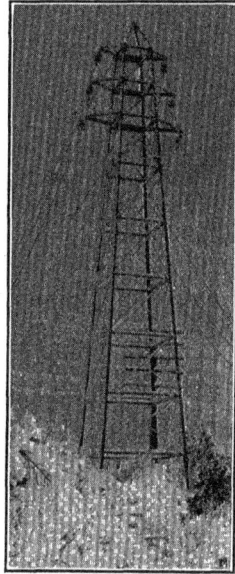

FIG. 20.

that the sleet remains in place throughout the span and that the wind rises to a velocity which in itself should occur but two or three times each year during the winter months. It has sometimes been specified that the thickness of sleet should be a factor of the diameter of the wire—an assumption which is not borne out by the facts. Indeed the effect of the sleet load is far greater upon small wires, since their area—and strength—is much smaller, while the wind and sleet loading is only a little less than for larger wires.

The following were reported, in 1915, as being the maximum loadings to which nine tower lines had actually been subjected in service, without injury. It should be noted that while the sleet load was usually measured, the wind load was not known and not reported.

No. 1.......... no sleet, 2 in. thickness of snow.
No. 2.......... 2 in. thickness of snow.
No. 3.......... $\frac{1}{4}$ in. thickness of sleet and $2\frac{1}{4}$ in. of soft snow.
No. 4.......... $\frac{5}{8}$ in. thickness of sleet.
No. 5.......... $\frac{3}{4}$ in. thickness of sleet—wind 35 miles.
No. 6.......... 1 in. thickness of sleet.
No. 7.......... 2 in. thickness of sleet.
No. 8.......... 2 cables parted during construction.
No. 9.......... 1 cable parted during construction.

The most reasonable assumption for general use in regions where sleet is known to occur would seem to be a thickness of $\frac{1}{2}$ in. all around the wires, combined with a wind load which will be discussed in the following section.

Wind.—It has been stated by some writers that it is necessary to know the probable direction of the wind and whether the wind and ice loads may occur together before the line may be designed.

In ordinary broken country and with the usual changes in direction of a line, the former information, even if obtainable, could hardly be of great service. For example, it would be very bad practice to assume that the wind would blow in but one direction, and to use a structure incapable of resisting pressure from the opposite direction.

The second condition mentioned contains a fallacious assumption in that it might be inferred that high winds and sleet will not occur together. For the vast majority of transmission lines in this country the sleet load, if there is ever sleet in the section in question, will probably occur during months in which high winds also occur. Again, it has frequently been claimed that high wind loads occurring during warm weather exert a greater effect upon the wires and their supports than the combined sleet and wind loads of the winter months. Disregarding cyclonic storms, which in many instances are beyond the limits of reasonable design, the writer believes the above claim to be entirely false and dangerously misleading. According to reports from the United States Weather Bureau, the maximum recorded wind pressures in many localities have occurred during the winter months. The combination of sleet and a moderately low wind velocity is greater in effect than the highest warm-weather wind pressures.

The Joint Report Specifications for crossings require a figured loading of ½ in. thickness of ice and 8 lb. per square foot wind pressure on the ice-covered diameter of the wires. This loading was considered by the framers of the specifications as being generally reasonable, and with the designated factor of safety,

Fig. 21.—Conductors deflected by high wind.

etc., to provide the proper construction for a crossing. It is not denied that thickness of ice greater than 0.5 in. or pressures of wind greater than 8 lb. may occur, but it is improbable that they will occur simultaneously over large areas or so frequently as to make it desirable to impose a greater loading on all future

3

crossings. All of the spans in a given line would never be subjected to the maximum figured load, nor would a number of adjacent short spans, or even one very long span, be likely to receive the maximum load over every lineal foot. It is true that telephone lines fail every winter, and perhaps that some old or incorrectly built low-voltage lines occasionally fall, but the writer, has yet to learn of the failure of a single wire strung, even approximately, to the Joint Report requirements. Further it should

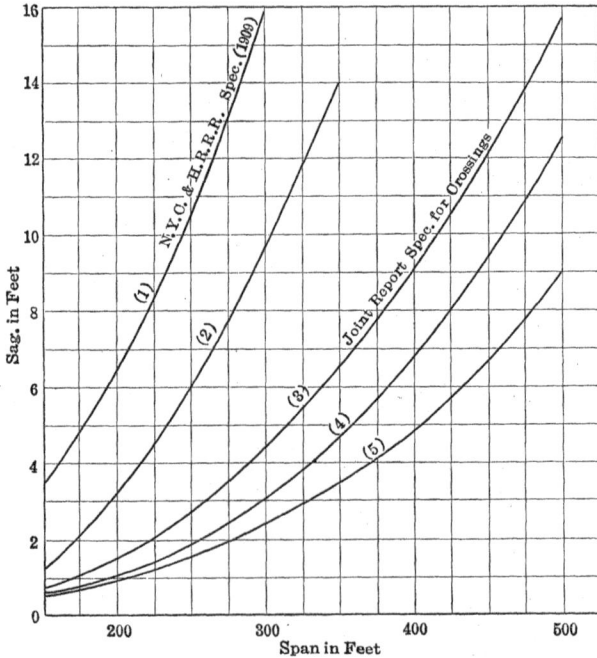

FIG. 22.—Comparative normal sags of No. 1 wire for various loadings.

be remembered that the loading and factors of safety in question —and these must be considered in conjunction—were recommended for crossings and for crossings only. As yet they have not been recommended by any authoritative body for general intermediate line construction.

Some specifications have provided for a load of 0.25 in. of ice and 8 lb. per square foot wind pressure with a stress limit of 0.9 of the elastic limit of the wire, while at least one crossing specification contains the severe requirement of 0.5 in. of ice and 20 lb. per square foot wind pressure with a stress limit of

0.4 of the ultimate strength of the wire. In order to indicate more clearly the relative effect of various loadings and factors of safety, the approximate curves in Fig. 22 have been prepared to show the normal sag (at 60°F. unloaded) of a No. 1, B. & S. gage, hard-drawn stranded copper wire under the following conditions:

(1)...0.5 in. ice + 20.0 lb. wind (120 miles per hour) max. stress, 0.4 ultimate = 1580 lb.
(2)...1.0 in. ice + 2.8 lb. wind (40 miles per hour) max. stress, 0.5 ultimate = 1980 lb.
(3)...0.5 in. ice + 8.0 lb. wind (70 miles per hour) max. stress, 0.5 ultimate = 1980 lb.
(4)...0.5 in. ice + 8.0 lb. wind (70 miles per hour) max. stress, 0.6 ultimate = 2370 lb.
(5)...0.25 in. ice + 8.0 lb. wind (70 miles per hour) max. stress, 0.9 elastic = 2110 lb.

The records of the United States Weather Bureau—omitting tornadoes, cyclones, and violent gales occurring in some particularly exposed localities—show a maximum indicated velocity of 100 miles per hour. The records at Bidston Observatory, Liverpool, England, covering the period from 1884 to 1888, give an actual velocity of 78 miles per hour as a maximum of 10 severe storms.[1]

Table 2 shows the maximum velocities observed at a number of stations by the United States Weather Bureau.

TABLE 2.—MAXIMUM WIND VELOCITIES

Observatory	Period	Maximum velocity indicated	Observatory	Period	Maximum velocity indicated
Chicago, Ill.	1871–1906	90	Savannah, Ga.	1894–1903	76
Buffalo, N. Y.	1871–1907	90	Philadelphia, Pa.	1872–1907	75
Galveston, Tex.	1894–1903	84	Bismarck, N. Dak.	1894–1903	72
New York, N. Y.	1871–1907	80	Boston, Mass.	1873–1907	72
Eastport, Me.	1873–1907	78	Salt Lake City, Utah.	1894–1903	60

Table 3 shows the three highest indicated velocities recorded each year by the United States Weather Bureau in its New York City station, during the period from 1884 to 1906 inclusive. This station was moved in March, 1895, from the Manhattan Life Insurance Building to the location at 100 Broadway; the latter is evidently in a more exposed position, as shown by the abrupt rise in velocities after 1895. The maximum velocity of 80 miles per hour occurred during a sleet storm.

[1] *Extract* from Overhead Construction for High-tension Electric Traction or Transmission, by R. D. Coombs, Transactions of American Society of Civil Engineers, Vol. LX.

TABLE 3.—RECORD OF HIGHEST WIND VELOCITIES IN NEW YORK CITY

Year	Date	Maximum velocity	Date	Maximum velocity	Date	Maximum velocity
1884	Oct. 18	44	Feb. 20	40	Dec. 9	40
5	Jan. 17	50	Dec. 7	50	Mar. 10	48
6	Feb. 26	64	Mar. 2	54	Jan. 9	44
7	Dec. 29	50	Nov. 16	48	Feb. 12	46
8	Jan. 26	60	Mar. 5	52	Mar. 13	50
9	Jan. 17	50	Feb. 1	48	Dec. 26	48
1890	Jan. 22	55	Dec. 17	48	Feb. 5	45
1	Dec. 30	53	Mar. 14	45	Jan. 11	44
2	Jan. 26	49	Mar. 11	40	Jan. 5	40
3	Aug. 29	54	Jan. 1	48	Oct. 13	48
4	Apr. 11	48	Oct. 10	48	Jan. 12	43
5	Dec. 27	73	Mar. 28	64	Aug. 4	62
6	Mar. 4	72	Feb. 7	65	Sept. 30	56
7	Jan. 18	60	Feb. 6	60	Oct. 17	60
8	Dec. 4	78	Sept. 7	72	Nov. 11	65
9	Mar. 20	80	Jan. 25	66	Feb. 27	64
1900	Oct. 16	76	Nov. 21	76	Jan. 26	76
1	Nov. 26	72	Jan. 19	72	Feb. 5	70
2	Mar. 19	74	Jan. 1	74	Feb. 2	74
3	July 2	72	Feb. 5	72	Sept. 17	65
4	Apr. 16	73	Sept. 15	68	Mar. 3	65
5	Dec. 10	64	Feb. 7	61	Apr. 10	56
6	Mar. 10	64	Jan. 6	61	Feb. 28	59

Table 4 is a record, by months, of the number of different 12-hour periods during which a maximum velocity of 60 miles, or more, was observed at the New York City station from 1895 to 1906 inclusive. Inasmuch as a maximum occurring late in one period and another early in the following period are both entered, a few of the entries represent the effects of the same storm.

For the vicinity of New York City, Tables 3 and 4 indicate that: the maximum velocities occur during the winter months, when sleet may be on the wires; indicated velocities of more than 80 miles per hour will rarely, if ever, occur during the life of a given structure; and indicated velocities of 65 to 75 miles per hour may be expected several times each year, though much less frequently in conjunction with sleet.

A rather complete study and tabulation of the U. S. Weather Bureau records of 43 observatories, located in 32 States, and

Table 4.—Number of 12-hour Periods in Which Wind Velocities of 60 Miles per Hour or Higher Were Observed in New York City

Month	Indicated velocities, in miles per hour																Totals
	60	61	62	63	64	65	66	67	68	70	72	73	74	76	78	80	
Jan	3	3	...	2	2	1	4	...	...	...	1	...	1	1	...	...	18
Feb	7	2	2	1	6	2	2	2	2	3	1	...	1	...	...	...	31
Mar	3	...	1	2	2	1	1	...	1	1	3	...	1	...	...	1	17
Apr	...	1	...	1	...	...	...	...	...	...	...	1	...	...	...	...	3
May	1	...	1	...	1	...	...	...	...	...	...	...	...	...	...	...	3
June	1	...	...	...	...	...	1	...	...	...	1	...	...	...	...	...	3
July	...	...	...	...	...	2	1	...	...	...	1	...	...	...	...	...	4
Aug	...	...	1	...	1	...	...	...	...	...	...	...	...	1	...	...	3
Sept	...	...	...	1	1	2	...	...	1	...	1	...	...	...	...	...	6
Oct	1	1	...	...	...	...	...	...	...	...	...	...	...	1	...	...	3
Nov	4	...	2	...	2	1	...	...	...	...	1	1	1	1	...	...	13
Dec	6	1	1	2	3	1	...	...	1	...	1	1	...	...	1	...	18
Totals	26	8	8	9	18	10	9	2	5	4	10	3	4	4	1	1	122

covering periods of observation from 5 to 43 years results in the following summary[1]:—

Total number of sleet storms........................ 487
Number with wind velocity over 40 miles per hour..... 31
Number with wind velocity over 50 miles per hour..... 12
Number with wind velocity over 60 miles per hour..... 5

Sleet formation, ¼ in. or less, during 90 storms
Sleet formation, ¼ in. to ½ in. during 62 storms
Sleet formation, ½ in. to 1 in. during 42 storms
Sleet formation, over 1 in. during 17 storms

Total recorded = 211

Temperature fell below 0° F., after sleet deposit ....3 storms
Maximum recorded wind velocity and maximum
sleet deposit occurring in same storm...........19 storms

Since the publication of Sir Isaac Newton's law for the pressures exerted by moving fluids—which, for wind pressures, may be reduced to the form

$$P = \frac{K}{370}V^2$$

in which P = pressure, in pounds per square foot, and V = velocity, in miles per hour—many investigators have experimented, with a view to the determination of values for the

[1] Handbook on Overhead Line Construction, National Elec. Light Assoc.

constant, K. For normal pressures against thin flat surfaces, most of the results indicate values between

$$P = 0.0035V^2 \qquad (1)$$

and
$$P = 0.0049V^2 \qquad (2)$$

These formulas, modified to apply to cylindrical surfaces, become

$$P = 0.0021V^2 \qquad (3)$$

and
$$P = 0.0029V^2 \qquad (4)$$

The Berlin-Zossen high-speed tests, in which wind pressures against trains were measured, gave the formula,

$$P = 0.0027V^2$$

and, using a rounded "nose" on the forward end,

$$P = 0.0025V^2$$

In Table 5 are given the equivalent actual velocities corresponding to those indicated by anemometer readings, and the pressures per square foot produced on flat and cylindrical surfaces.

TABLE 5.—WIND PRESSURES AND VELOCITIES CORRESPONDING TO ANEMOMETER READINGS

Indicated velocities, mi. per hr.	Actual velocities, mi. per hr.	Pressure per sq. ft. on cylinders, $P = 0.0025V^2$	Pressure per sq. ft. on flat surfaces, $P = 0.0042V^2$
30	25.7	1.7	2.8
40	33.3	2.8	4.6
50	40.8	4.2	7.0
60	48.0	5.8	9.7
70	55.2	7.6	12.8
80	62.2	9.7	16.2
90	69.2	12.0	20.1
100	76.2	14.6	23.3
110	83.2	17.3	29.1
120	90.2	20.3	34.2

P = pressure, in pounds per square foot.
V = velocity (actual), in miles per hour.

Assuming an indicated velocity of 70 miles per hour, or an actual velocity of 55.2 miles per hour, the above equation for obtaining pressures against flat surfaces becomes $P = 12.8$ lb. per square foot of projected area, while for pressures against cylindrical surfaces $P = 7.6$ lb. per square foot of projected area.

On long spans, the maximum pressure at one point may be considerably in excess of the equivalent uniform pressure along the wire, while very short spans may be exposed to the maximum pressure throughout their length. In view of the rare, if not improbable, occurrence of indicated velocities greater than 80 miles per hour, and the further improbability of such winds accompanying sleet storms, or of the sleet remaining in place, the following pressures seem to be reasonable for general use:

$P = 13.0$ lb. per square foot of projected area for flat surfaces;
$P = 8.0$ lb. per square foot of projected area of wires covered with 0.5-in. deposit of ice.

In applying wind loads to the supports, wooden, concrete or cylindrical metal poles should be considered as flat surfaces. By so doing the excess loading will compensate for the increased surface at the top of the poles due to arms, braces, insulators, etc. Latticed steel poles or steel towers should be treated as having flat surfaces equal to the exposed area of the members on the windward side, increased by 50 per cent. to allow for pressure on the leeward side of the poles and by 100 per cent. for wide-base towers.

On the other hand some decreases in pressure would be justified on the lower part of poles or towers, except when set on hill-tops.

Broken Wires.—In determining the proper wire loads and factors of safety to be used, it is important to bear in mind the effect of any further requirement such as a provision for dead-ending or carrying broken wires, inasmuch as the effect of the latter requirement is to impose from 5 to 40 times the former loading upon the insulator connections and the supporting structures. Dead-ending and corner-turning are different only in degree, and designing for a broken-wire load is more or less equivalent to designing all structures as corner structures.

Fig. 23 is a graphical representation of the relative effect of what is termed balanced transverse loading and a broken-wire condition. The ordinates are the ratios of loads caused by one broken wire to the loads caused by balanced spans. In other words, the ordinates show how many times more severe a broken-wire condition is than the load of the same wire unbroken under identical ice and wind loads. For instance, a No. 1 hard-drawn stranded copper wire in spans of 200 ft. would, for a broken-wire condition, impose 16 times the stress upon its support that it would unbroken.

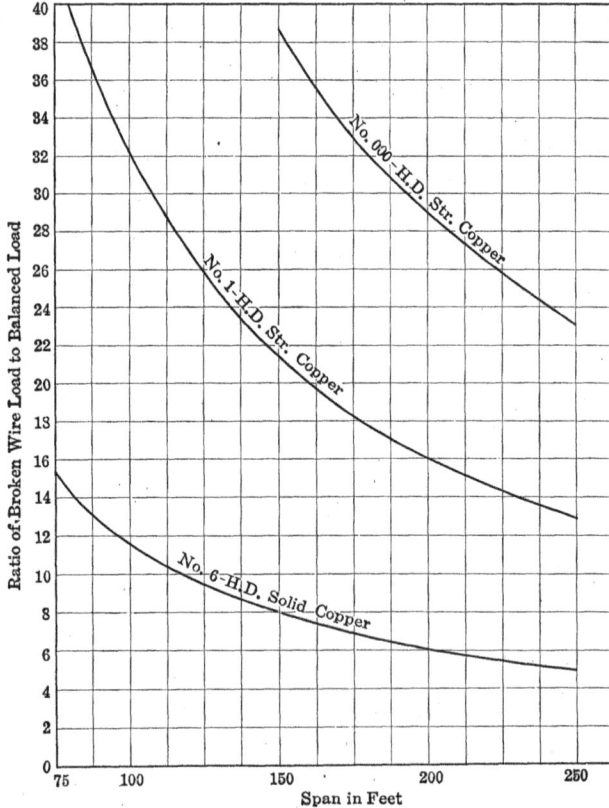

Fig. 23.—Relative effect of balanced and broken wire loads.

In determining a mechanical factor of safety for any material it is customary, whether so stated or not, to assume certain portions of the factor as safeguarding each of the possible elements of danger, such as errors of design, workmanship, excess loads and deterioration of material. When using wire cables a relatively low factor may be assumed, since a wire catenary, both in material and as a structural member, is more uniform in section, strength and elasticity, and less influenced by eccentric

loads or errors of workmanship, than any other engineering structure. Therefore, failure in the wires may be considered as resulting usually from electrical causes such as
arcs in the span or at the insulators. Assuming the provision
of adequate clearance and proper spacing of wires in the span,
the majority of wire troubles should occur at the insulators.
Further, and in view of the small number of failures per insulator in the existing installations, it would seem that the increasing tendency to improve the insulation should have some
effect in lowering the number of broken wires per support.

In consideration of the above it is, in the writer's opinion at
least, utterly indefensible to assume a severe broken-wire condition in designing all poles and towers. Particularly is this true
if future construction must actually accord with specifications.
There can be no engineering justification for a specification
which premises a large proportion of wires broken under full
load, when the devices fastening the conductors to the supports would not withstand any considerable part of such a load.
Again, it is probably a fact that many of the structures in existing lines are not as strong as the preliminary test structure. This
may be due to a variety of causes, such as local injury or incipient bends in light sections, lack of rigidity in the foundations, weakness in torsion, and last but not least, the usual
difference between test specimens and the least perfect field
product.

In working backward from the results of practical experience over large areas, the tendency is to overestimate both the
actual loading and the strength of the structures. In other
words, many existing lines, particularly the heavy wooden pole
lines, remain in service without failing not because they have a
strength equivalent to some recent requirements, but simply
because they have never been subjected to such loads. Therefore if a severe mechanical requirement is placed in a standard
specification it must be assumed that designers will eventually
be driven to literal compliance therewith, and the net result will
not be equivalent to the designs of the transition period upon
which the requirement is supposed to be founded.

The assumed load on the wires should equal as nearly as
possible the maximum load that may be expected on some
indeterminate number of spans during some indeterminate
interval of time. One or more spans in a given line may con-

ceivably, and properly, receive a greater load during their life-time. Such excess loads may, or may not, be harmful, depending on the factor of safety. A large factor of safety will undoubtedly continue to protect inaccurate assumptions of loading, but the use of unreasonable loads and impossible stresses does not establish wise engineering standards.

Bridges and buildings are not designed to withstand tornadoes, nor need power wires be absolutely immune from failure. It is, however, becoming more and more important to provide continuous service and to establish a standard which will satisfy all conflicting interests without unduly burdening a great industry.

The difficulty in specifying a sliding scale of broken wires is that in reality only one or two wires may be reasonably expected to break, whether there is one circuit upon the structure or one dozen circuits. If there are only three wires on a pole, the load requirement of one broken wire is relatively greater than the requirement of several broken wires on a structure of many wires on account of the pullback of adjoining spans.

Again, if broken wires are to be considered, the wire connections must be designed to withstand a broken-wire loading, otherwise the broken-wire load could not be transmitted to the support.

To allow pullback in a consistently designed line is correct both theoretically and practically, but its accurate computation is quite difficult, and the inclusion of such a condition in a general specification is probably inadvisable. It appears, therefore, that a broken-wire load should be applied to the arms and wire connections, but that its application to the supports may depend in a measure upon the character of the supports.

CHAPTER III

WIRES AND CABLES

The qualities desired in electric-service wires, in so far as the construction is concerned, are mechanical strength, tenacity, and ability to resist corrosion or other deterioration. In ordinary practice, the breaking strength required for a wire of a given span will depend entirely on the sag, because increasing the sag will reduce the wire tension approximately in proportion to the sag. Practical considerations, however, indicate a rather indefinite minimum, below which it is undesirable to go. Any surface injury, such as local pitting by arcs and nicks caused by careless handling, or any weakness in the material due to errors in manufacture, will have relatively greater effect on a small wire than on a large one. Moreover, such faults are more serious in solid wires than in stranded cables, and in hard-drawn wire than in soft-drawn wire. In stranded cables, an injury to a single strand affects only a fractional part of the entire section; in hard-drawn wire the surface, or skin material, has approximately twice the unit strength of the interior mass, so that an injury will have a relatively greater effect on hard-drawn wire. Fortunately, however, the process of wire drawing insures a large amount of work per unit of mass, so that the finished product is a very homogeneous and trustworthy material. This quality, combined with the reduction in stress resulting from any increase in sag caused by stretching, explains the comparative immunity from mechanical failures.

Copper.—The good qualities of copper wire are a matter of common knowledge, and as stated previously, its manufacture and method of use combine to make it almost unique as a material of construction. It is fairly immune from corrosive action, as ordinarily used in transmission-line work, although it is not absolutely indestructible. The principal sources of injury to copper are due to its softness and low melting point. The former renders it liable to nicks or broken strands in stringing and clamping, and the latter to burning by arcs.

43

Unless the voltage is such that insulated or weatherproof wire affords some real protection, there is no logical structural reason for using it. Otherwise, it merely serves as an additional load and offers a greater diameter for sleet deposits, besides deteriorating far in advance of the rest of the construction and frequently hanging in unsightly streamers.

Since the sheen of freshly strung copper is greatly lessened after exposure, it becomes problematical whether the attention of the casual observer would in reality be attracted more by the copper or by the size and spacing of insulators which cannot be disguised.

Copper Covered.—A comparatively recent development in transmission-line wires is the use of a steel wire covered with copper. This product is produced by drawing out an ingot of steel which has been previously encased in a copper covering. The thickness of the shell of copper may be varied within wide limits, the usual commercial proportions being an amount of copper that, combined with the lower conductivity of the steel core, produces a wire having either 30 or 40 per cent. of the conductivity of a copper wire of the combined gage. Thirty per cent. copper-covered wire is about 5 per cent. stronger than 40 per cent. wire of the same gage. The thickness of the shell of copper is quite small, depending in part on the size of the wire. Such wire should, therefore, be handled at least as carefully as copper wire. Since the thickness of the copper decreases with the size of the wires in the cable, it is preferable, at least for overhead ground wires, to use the 40 per cent. grade or else to use cables of few strands.

The steel from which the wire is drawn is a high-carbon steel having an ultimate strength of about 90,000 lb. per square inch, and a correspondingly high elastic limit. During the process of copper coating and wire drawing, there is an annealing effect followed by hardening, the net result being to produce a wire having an ultimate strength not greatly below that of the original ingot material. In general, and with the grade of steel commonly used by manufacturing companies, the ultimate strength of copper-covered wire is from 20 per cent. to 40 per cent. greater than that of the corresponding sizes of hard-drawn copper.

The principal uses of this material in transmission work are for overhead ground wires, telephone wires, and the power wires of the lighter and lower capacity lines, where little future growth of business may be expected.

Aluminum.—Aluminum wire, as now used for power-line purposes, is usually employed in the form of stranded cables, and when so used is no longer subject to some of the troubles incident to the earlier installations. As a material, it is quite different from copper, although used for similar purposes. Therefore, in making price comparisons, it is necessary to consider not only the price per mile per unit of electrical rating, but also the changes in the general construction of the line.

The conductivity of aluminum is about 60 per cent., based on the Matthiesen standard for copper, making aluminum cables about 1.5 times the area and 1.25 times the diameter of copper cables having equal conductivity. As the specific weight of aluminum is about 0.33 that of copper, the weight of aluminum cable will be about 0.5 times that of copper cable having the same conductivity.

The strength of aluminum is about 0.8 that of soft copper and 0.4 the strength of hard copper.

The net result of these differences is best shown by a concrete example. Thus, a No. 00 aluminum cable has about the same conductivity as a No. 1 stranded hard-drawn copper cable, and their other characteristics for a 400-ft. span are as follows:

	No. 00 aluminum	No. 1 copper
Breaking strength, pounds............................	2500	3600
Elastic limit, pounds...............................	1460	2180
Wind pressure on ice-covered diam., lb. per foot...	0.943	0.885
Weight of ice-covered cable, lb. per foot..........	0.691	0.770
Resultant load, lb. per foot.......................	1.168	1.173
Maximum wire tension (factor 2.0), pounds.......	1250	1800
Transverse load, per wire, pounds...............	377	354
Normal sag.....................................	20 ft.	10 ft.
Maximum sag...................................	21 ft.	12 ft.

Since the coefficient of expansion of aluminum is considerably higher than that of copper, aluminum cables are more affected by temperature changes. Relatively greater sags will therefore occur in hot weather, while at low temperatures there is a greater increase in tension due to the contraction of the material with its resultant decrease in sag. In addition, the lighter weight of aluminum renders it more liable to local displacement by wind pressure. It is necessary, as a result of these differences in the material, to provide greater pin separation and overhead clearance for aluminum conductors.

From a construction viewpoint, however, the lighter weight of aluminum does not possess any particular merit, except possibly greater ease of handling the reels and pulling out wire in stringing. The saving in dead load on the supports is negligible and is more than offset by the greater sag and separation required. Furthermore, the increased diameter imposes a greater wind load. The choice between aluminum and copper will depend on the relative cost of the two materials and their accompanying construction, together with the allowances which can be made for the scrap values of the two installations.

Steel.—Steel cable can be obtained of almost any desired unit breaking strength, the commercial grades ranging from the low grade steel of guy wire, which has an ultimate strength of 60,000 lb. per square inch, to steels of 200,000 lb. or more.

For transmission-line purposes steel cable is used chiefly for overhead ground wires or as power wires for very long spans. The occasional use of steel messengers for telephone or insulated cables does not involve any considerable quantity of such material employed on typical transmission lines.

Steel cables for line work should always be galvanized, and should be larger than is actually required for strength.

The galvanizing of cable is by no means as permanent a protection as the hot-dip, unwiped process applied to structural steel; therefore some allowance should be made for future corrosion.

Despite the temptation to use small-gage cables made of the higher grade steels, on account of their greater strength, it is generally preferable to adhere to medium grades such as the Siemens-Martin. Guy-strand steel cable is the lowest commercial grade and its quality is relatively much lower than that of any of the higher grades. Large diameter cables, particularly of high-grade steel, are rather difficult to handle as they are very stiff. It should be noted that all cables of great strength require special clamping attachments for dead ending, the ordinary quota of clips and clamps being inadequate to transmit the tension.

Telephone Wire.—Supporting a telephone circuit on long-span transmission-line structures introduces a difficulty, in that the small wires which are sufficient for telephone service do not have the mechanical strength to carry safely in long spans.

It is practicably impossible to string any ordinary telephone wires so that they will be reasonably secure on long-span lines. There have arisen, therefore, two general methods of procedure; one is to use larger and stronger wires; and the other to contemplate failure in the telephone circuit as a necessary evil. Solid steel wire No. 6 BWG, sometimes called river crossing wire, has an ultimate strength about equal to that of No. 00 stranded hard-drawn copper and can, therefore, be strung equally well in long-span lines.

Catenary.—If two ends of an imaginary wire having perfect flexibility and uniformity of material but no ductility were supported at two points in the same horizontal plane the wire would take the shape of a curve known as the catenary. For all practical purposes it may be assumed that actual wires and cables possess the same characteristics. The curve of the wires between the supports is therefore known, if the span and sag are known. Inasmuch, however, as the equation of the catenary is rather complicated, while that of the parabola, which closely resembles it, is simple, the latter is usually employed instead; thus,

$$y^2 = ax$$

"a" being a constant found by substituting the known value of x for the point on the curve at the support, *i.e.*,

$a = \dfrac{S^2}{4d}$ in which S is the length of the span and d is the sag.

Assume a uniform load on each lineal foot of the span, and imagine half of the span removed and the wire held in place by the tension T at the middle of the span. Then considering the moments about the remaining support we have the weight of the half span multiplied by its lever arm which is one-quarter of the span, equals the balancing force T multiplied by its lever arm d.

Since the total weight W = the weight per foot, $\times$ the half span or $W = \dfrac{ws}{2}$

we have

$$\frac{wS}{2} \times \frac{S}{4} = Td$$

Therefore

$$\frac{wS^2}{8} = Td$$

and

$$T = \frac{wS^2}{8d}$$

or, the tension in the wire equals the weight per foot times the span squared divided by eight times the sag.

It is necessary, however, to take into consideration the effect of a change in length of the wire due to temperature and loading, and a simple arrangement of formulæ in which this is done is given below.

The following mathematical treatment is not new,[1] but the writer has found the arrangement convenient:

S = span, in feet.

d = sag, in feet.

W = load per lineal foot in plane of wire.

A = area of wire, in square inches.

E = modulus of elasticity.

θ = coefficient of expansion.

t = change of temperature, in degrees.

e = elongation or change of length, within elastic limit.

L_o = length, in feet, of imaginary wire ($W = 0$) at normal temperature.

L_{oc} = length, in feet, of imaginary wire, cold ($t°$F. below normal temperature).

L_{oh} = length, in feet, of imaginary wire, hot ($t°$F. above normal temperature).

Index to Subscripts.—

No subscript = normal conditions.

　　　$_c$ = cold: $t°$F. below normal + dead load.

　　　$_i$ = cold: ice load + dead load.

　　　$_{cw}$ = cold: wind load + dead load.

　　　$_{iw}$ = cold: ice + wind + dead load.

　　　$_h$ = hot: $t°$F. above normal + dead load.

W_{iw} is the resultant of the vertical dead + ice loads and the horizontal wind load.

W_{cw} is the resultant of the vertical dead load and the horizontal wind load.

---

[1] Overhead Construction for High-tension Electric Traction or Transmission, by R. D. Coombs, Trans. Am. Soc. C. E., Vol. LX (1908).

Stresses.—Substitute normal values in Eqs. 1, 2, 3, and 4. Assume values of T_h, T_{iw}, T_c, T_i, or T_{cw}, such that Eqs. 5 and 6 will give identical values of d_h, d_{iw}, etc. The tension that will give the same sag by Eqs. 5 and 6 (independently) is the tension resulting from that sag and the given loading.

$$T = \frac{WS^2}{8d} \tag{1}$$

$$L = S\left[1 + \frac{8d^2}{3S^2}\right] \tag{2}$$

$$e = \frac{TL}{EA} \tag{3}$$

$$L_o = L - e \tag{4}$$

(*t°F. above normal, with dead load.*)

$$L_{oh} = L_o(1 + \theta t_h) \quad e_h = \frac{L_{oh} \times T_h}{EA} \quad L_h = L_{oh} + e_h$$

$$d_h = 0.612\sqrt{S\,(L_h - S)} \tag{5}$$

$$d_h = \frac{W_h \times S^2}{8T_h} \tag{6}$$

(*t°F. below normal, with dead + ice + wind loads.*)

$$L_{oc} = L_o(1 - \theta t_c) \quad e_{iw} = \frac{L_{oc} \times T_{iw}}{EA} \quad L_{iw} = L_{oc} + e_{iw}$$

$$d_{iw} = 0.612\sqrt{S(L_{iw} - S)} \tag{5}$$

$$d_{iw} = \frac{W_{iw} \times S^2}{8T_{iw}} \tag{6}$$

(*t°F. below normal, with dead load.*)

$$L_{oc} = L_o(1 - \theta t_c) \quad e_c = \frac{L_{oc} \times T_c}{EA} \quad L_c = L_{oc} + e_c$$

$$d_c = 0.612\sqrt{S(L_c - S)} \tag{5}$$

$$d_c = \frac{W_c \times S^2}{8T_c} \tag{6}$$

(*t°F. below normal, with dead + ice loads.*)

$$L_{oc} = L_o(1 - \theta t_c) \quad e_i = \frac{L_{oc} \times T_i}{EA} \quad L_i = L_{oc} + e_i$$

$$d_i = 0.612\sqrt{S(L_i - S)} \tag{5}$$

$$d_i = \frac{W_i \times S^2}{8T_i} \tag{6}$$

4

(*t°F. below normal, with dead + wind loads.*)

$$L_{oc} = L_o(1 - \theta t_c) \qquad e_{cw} = \frac{L_{oc} \times T_{cw}}{EA} \qquad L_{cw} = L_{oc} + e_{cw}$$

$$d_{cw} = 0.612\sqrt{S(L_{cw} - S)} \tag{5}$$

$$d_{cw} = \frac{W_{cw} \times S^2}{8T_{cw}} \tag{6}$$

In Tables 6 to 13 are given the physical properties and the wind and ice loads for various wire gages.*

TABLE 6.—PROPERTIES OF WIRE MATERIAL

	Ultimate strength, lb. per square inch	Elastic limit, lb. per square inch[1]	Modulus of elasticity, E	Coefficient of expansion, per °F.
Copper, solid soft-drawn........	32 to 34,000	16,000	14,000,000	0.0000096
Copper, solid med.-drawn.......	40 to 50,000	22 to 27,000	15,000,000	0.0000096
Copper, solid hard-drawn........	50 to 60,000	30 to 35,000	16,000,000	0.0000096
Copper, strand soft-drawn.......		30,000	15,000	8,000,000	0.0000096
Copper, strand med.-drawn......		45,000	25,000	10,000,000	0.0000096
Copper, strand hard-drawn.....		55,000	33,000	12,000,000	0.0000096
Copper clad, solid, hard-drawn...	60 to 90,000	35 to 53,000	21,000,000	0.0000067
Copper clad, strand hard-drawn..	70 to 90,000	41 to 53,000	18,000,000	0.0000067
Aluminum, strand.............	23 to 24,000	14,000	9,000,000	0.0000128
Steel strand, Siemens-Martin....	75,000	....	25,000,000	0.0000066
Steel strand, high-strength.......	150,000	....	25,000,000	0.0000066
Steel strand, ex-high-strength....	180,000	....	25,000,000	0.0000066
Steel solid ex-high-strength.....	187,000	....	29,000,000	0.0000066

It has been urged by some that using the 0.5-in. ice and 8-lb. wind load with the parabolic formula for computing the stress in the wire does not give results which accord with experience, since actual spans erected with less than the specified sags have not failed in service. On the other hand, it is sometimes claimed that allowing maximum stresses near the elastic limit is dangerous. The facts of the matter are:

First, the true catenary formula is scientifically and mathematically correct within the elastic limit. *Second*, the parabolic formula, ordinarily used for simplicity, gives results which in the vast majority of cases are closer to the exact values than the actual wire stringing will be to the specified stringing. *Third*, the material of a wire catenary is more uniform in section,

*Tables from R. D. Coombs & Co. design standards.

[1] The elastic limit used is in reality the yield point, or point of appreciable extension, as this value seems more applicable to wire stringing than that obtained by accurate laboratory tests.

strength, and other characteristics than that of any other engineering structure; therefore the error of design is correspondingly less. *Fourth*, the stretch of a ductile material, such as copper, permits the sag to increase and the stress to decrease and, within limits, does not perceptibly decrease the cross-section at any

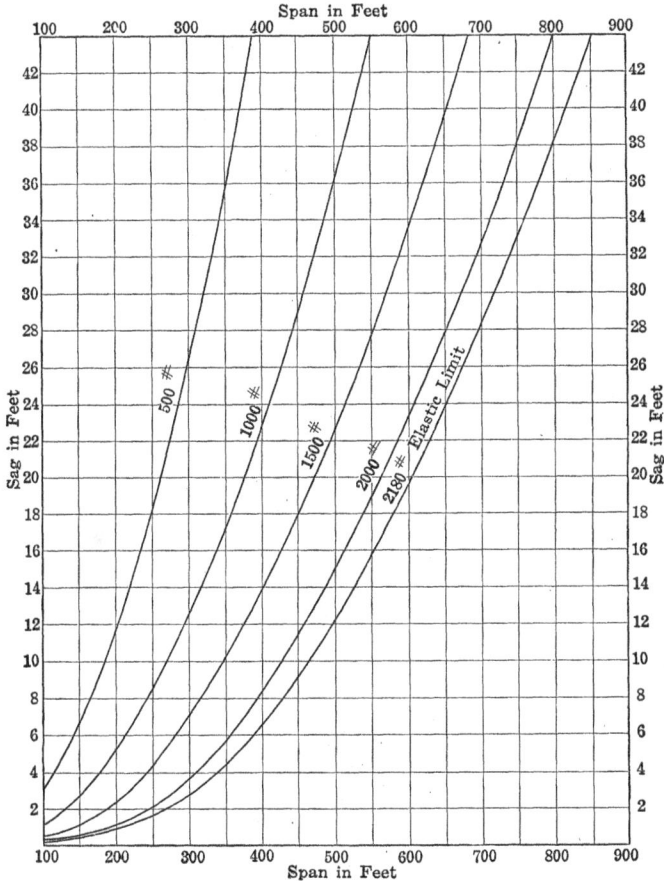

FIG. 24.—Sags and tensions of No. 1 H. D. copper.

point. Therefore, a loaded span stretches enough to relieve the stress and does not fail unless the load is very excessive.

Fifth, the specified maximum loading is an emergency loading and is not a general or frequent occurrence on any span or line. The facts of the matter are that the spans in service

have either not been subjected to loads in excess of those required by the catenary formula to develop their elastic limit,
or the wire has stretched and the sag has increased. Possibly there are two other reasons for seemingly overstressed lines
giving satisfactory service. One is that the poles have bent

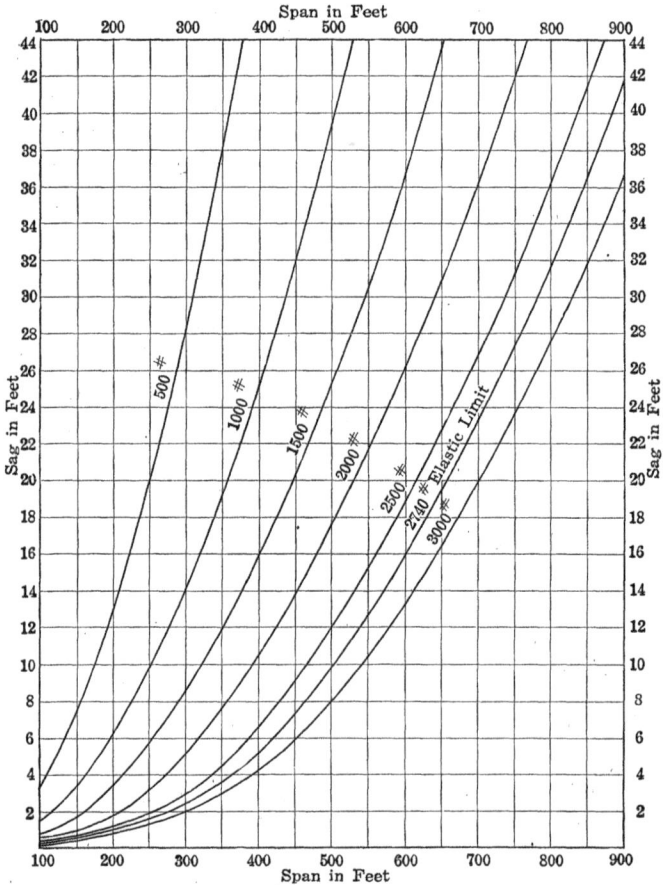

Fig. 25.—Sags and tensions of No. 0 H. D. copper.

or the wires slipped through the ties, thus temporarily increasing
the sag. Second, the wires may have become tempered or
hardened by tension, possibly by atmospheric changes or other
action, and by becoming harder have been able to sustain a
greater load.

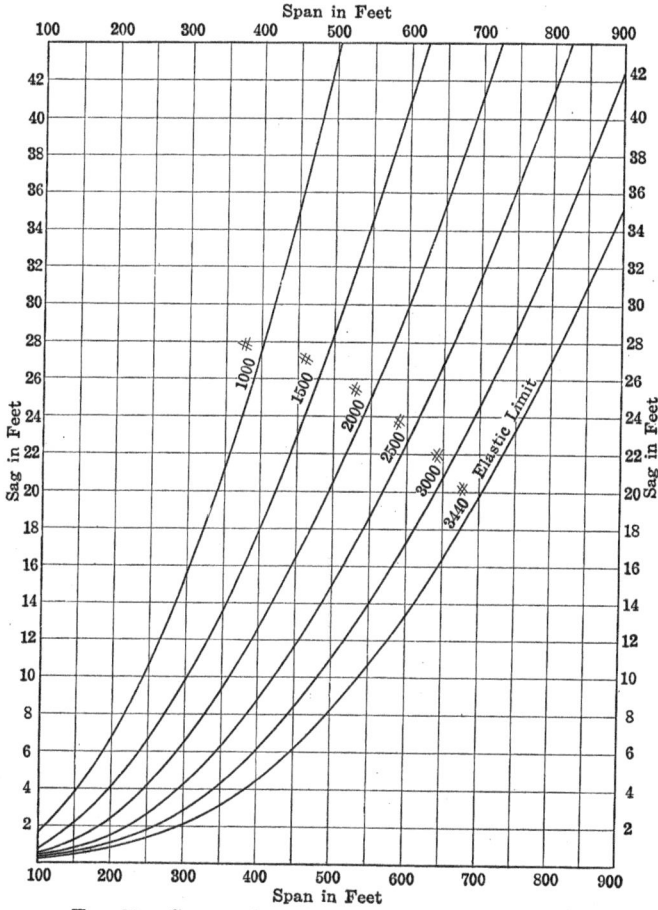

Fig. 26.—Sags and tensions of No. 00 H. D. copper.

TABLE 7.—PROPERTIES OF BARE STRANDED COPPER CABLE

| Gage, B. & S. or circ. mils | Diameter, inches | Area, sq. in. | Ultimate strength | | | Elastic limit | | | Load per lineal foot | | | | | | | | | EA | | |
| | | | Hard, 55,000 | Med., 45,000 | Soft, 30,000 | Hard, 33,000 | Med., 25,000 | Soft, 15,000 | Vertical | | | Horizontal | | | Plane of resultant | | | Hard, E = 12,000,000 | Medium, E = 10,000,000 | Soft, E = 8,000,000 |
									Dead	Dead + 0.5-in. ice	Dead + 0.75-in. ice	15.0 lb. per sq. ft.	8.0-lb. + 0.5-in. ice	11.0-lb. + 0.75-in. ice	Load A	Load B	Load C			
2,000,000	1.630	1.5687	86,280	70,590	47,060	51,770	39,220	23,530	6.205	7.530			1.753			7.731		18,824,400	15,687,000	12,549,600
1,750,000	1.526	1.3649	75,070	61,420	40,950	45,040	34,120	20,470	5.429	6.691			1.684			6.900		16,378,800	13,649,000	10,919,200
1,500,000	1.412	1.1783	64,810	53,020	35,350	38,880	29,460	17,670	4.654	5.843			1.608			6.060		14,139,600	11,783,000	9,426,400
1,250,000	1.289	0.9817	53,990	44,180	29,450	32,400	24,540	14,730	3.878	4.991			1.526			5.219		11,780,400	9,817,000	7,853,600
1,000,000	1.152	0.7849	43,170	35,320	23,550	25,900	19,620	11,770	3.100	4.128			1.435			4.370		9,418,800	7,849,000	6,279,200
750,000	0.998	0.5892	32,410	26,510	17,680	19,440	14,730	8,840	2.325	3.257			1.332			3.519		7,070,400	5,892,000	4,713,000
500,000	0.813	0.3924	21,580	17,660	11,770	12,950	9,810	5,890	1.548	2.366			1.209			2.657		4,708,800	3,924,000	3,139,200
450,000	0.772	0.3523	19,380	15,850	10,570	11,630	8,810	5,280	1.393	2.184			1.181			2.483		4,227,600	3,523,000	2,818,400
400,000	0.728	0.3143	17,290	14,140	9,430	10,370	7,860	4,710	1.239	2.003			1.152			2.311		3,771,600	3,143,000	2,514,400
350,000	0.679	0.2751	15,130	12,380	8,250	9,080	6,880	4,130	1.083	1.817			1.119			2.134		3,301,200	2,751,000	2,200,800
300,000	0.629	0.2359	12,970	10,620	7,080	7,780	5,900	3,540	0.926	1.628			1.086			1.956		2,830,800	2,359,000	1,887,200
250,000	0.574	0.1964	10,800	8,840	5,890	6,480	4,910	2,950	0.772	1.440			1.049			1.782		2,356,800	1,964,000	1,571,200
0000	0.528	0.1661	9,140	7,470	4,980	5,480	4,150	2,490	0.653	1.293			1.019			1.646		1,993,200	1,661,000	1,328,800
000	0.464	0.1319	7,250	5,940	3,960	4,350	3,300	1,980	0.512	1.112			0.976			1.480		1,582,800	1,319,000	1,055,200
00	0.413	0.1043	5,740	4,690	3,130	3,440	2,610	1,560	0.407	0.975			0.942			1.356		1,251,600	1,043,000	834,400
0	0.368	0.0830	4,560	3,730	2,490	2,740	2,080	1,250	0.322	0.862			0.912			1.255		996,000	830,000	664,000
1	0.328	0.0660	3,630	2,970	1,980	2,180	1,650	990	0.255	0.770			0.885			1.173		792,000	660,000	528,000
2	0.292	0.0520	2,860	2,340	1,560	1,720	1,300	780	0.203	0.695			0.861			1.106		624,000	520,000	416,000
3	0.260	0.0413	2,270	1,860	1,240	1,360	1,030	620	0.161	0.634			0.840			1.052		495,600	413,000	330,400
4	0.232	0.0328	1,800	1,480	980	1,080	820	490	0.127	0.582			0.821			1.006		393,600	328,000	262,400
5	0.206	0.0260	1,430	1,170	780	860	650	390	0.101	0.540			0.804			0.969		312,000	260,000	208,000
6	0.184	0.0206	1,130	930	620	680	515	310	0.080	0.505			0.789			0.937		247,200	206,000	164,800

NOTE.—A Loading = dead load and 15.0 lb. per square foot wind pressure.

B Loading = dead + 0.5-in. ice and 8.0-lb. wind pressure.

C Loading = dead + 0.75-in. ice and 11.0-lb. wind pressure.

TABLE 8.—PROPERTIES OF SOLID COPPER WIRE

Gage, B.&S.	Diameter, inches	Area, sq.in.	Ultimate strength Hard, 50,000/60,000	Ultimate strength Med., 40,000/50,000	Ultimate strength Soft, 32,000/34,000	Elastic limit Hard, 30,000/35,000	Elastic limit Med., 22,000/27,000	Elastic limit Soft, 16,000	Vertical Dead	Vertical Dead + 0.5-in. ice	Vertical Dead + 0.75-in. ice	Horizontal 15.0 lb. per sq.ft.	Horizontal 8.0-lb. + 0.5-in. ice	Horizontal 11.0-lb. + 0.75-in. ice	Resultant Load A	Resultant Load B	Resultant Load C	EA Hard, E = 16,000,000	EA Medium, E = 15,000,000	EA Soft, E = 14,000,000
0000	0.460	0.1662	8,310	6,650	5,315			2,660	0.641	1.238			0.973			1.575		2,659,200	2,493,000	2,326,800
000	0.410	0.1318	6,590	5,270	4,220			2,110	0.509	1.074			0.940			1.427		2,108,800	1,977,000	1,845,200
00	0.365	0.1045	5,225	4,180	3,340			1,670	0.403	0.940			0.910			1.309		1,672,000	1,567,500	1,463,000
0	0.325	0.0829	4,560	3,730	2,740			1,325	0.320	0.833			0.883			1.214		1,326,400	1,243,500	1,160,600
1	0.289	0.0657	3,745	3,090	2,170			1,050	0.253	0.744			0.860			1.137		1,051,200	985,500	919,800
2	0.258	0.0521	3,125	2,605	1,770			835	0.202	0.673			0.838			1.075		833,600	781,500	729,400
3	0.229	0.0413	2,480	2,065	1,405			660	0.159	0.613			0.820			1.024		660,800	619,500	578,200
4	0.204	0.0328	1,970	1,640	1,115			525	0.126	0.564			0.803			0.981		524,800	492,000	459,200
5	0.182	0.0260	1,560	1,300	885			415	0.100	0.524			0.788			0.946		416,000	390,000	364,000
6	0.162	0.0206	1,235	1,030	700			330	0.079	0.491			0.775			0.917		329,600	309,000	288,400

TABLE 9.—PROPERTIES OF SOLID, TRIPLE-BRAIDED WEATHERPROOF COPPER WIRE

Gage, B.&S.	Diameter, inches	Area, sq.in.	Ultimate strength Hard, 50,000/60,000	Ultimate strength Med., 40,000/50,000	Ultimate strength Soft, 32,000/34,000	Elastic limit Hard, 30,000/35,000	Elastic limit Med., 22,000/27,000	Elastic limit Soft, 16,000	Vertical Dead	Vertical Dead + 0.5-in. ice	Vertical Dead + 0.75-in. ice	Horizontal 15.0 lb. per sq.ft.	Horizontal 8.0-lb. + 0.5-in. ice	Horizontal 11.0-lb. + 0.75-in. ice	Resultant Load A	Resultant Load B	Resultant Load C	EA Hard, E = 16,000,000	EA Medium, E = 15,000,000	EA Soft, E = 14,000,000
0000	0.640	0.1662	8,310	6,650	5,315			2,660	0.767	1.476			1.093			1.837		2,659,200	2,493,000	2,326,800
000	0.593	0.1318	6,590	5,270	4,220			2,120	0.629	1.309			1.062			1.686		2,108,800	1,977,000	1,845,200
00	0.515	0.1045	5,225	4,180	3,340			1,670	0.502	1.133			1.010			1.518		1,672,000	1,567,500	1,463,000
0	0.500	0.0829	4,560	3,730	2,740			1,325	0.407	1.029			1.000			1.434		1,326,400	1,243,500	1,160,600
1	0.453	0.0657	3,745	3,090	2,170			1,050	0.316	0.909			0.968			1.328		1,051,200	985,500	919,800
2	0.437	0.0521	3,125	2,605	1,770			835	0.260	0.843			0.958			1.276		833,600	781,500	729,400
3	0.406	0.0413	2,480	2,065	1,405			660	0.199	0.763			0.937			1.208		660,800	619,500	578,200
4	0.359	0.0328	1,970	1,640	1,115			525	0.164	0.698			0.906			1.143		524,800	492,000	459,200
5	0.344	0.0260	1,560	1,300	885			415	0.135	0.660			0.896			1.113		416,000	390,000	364,000
6	0.328	0.0206	1,235	1,030	700			330	0.112	0.627			0.885			1.084		329,600	309,000	288,400

NOTE.—A Loading = dead load and 15.0 lb. per square foot wind pressure.
B Loading = dead + 0.5-in. ice and 8.0-lb. wind pressure.
C Loading = dead + 0.75-in. ice and 11.0-lb. wind pressure.

TABLE 10.—PROPERTIES OF STRANDED ALUMINUM CABLE

| Gage, B. & S. | Diam., in. | Area, sq. in. | Ultimate strength | Elastic limit | Load per lineal foot | | | | | | | | | EA |
| | | | | | Vertical | | | Horizontal | | | Resultant | | | |
			23,000 24,000	14,000	Dead	Dead + 0.5-in. ice	Dead + 0.75-in. ice	15.0 lb. per sq. ft.	8.0-lb. + 0.5-in. ice	11.0-lb. + 0.75-in. ice	A	B	C	E = 9,000,000
500,000	0.814	0.3924	9,025	5,500	0.460	1.280	......	......	1.209	......	......	1.762	......	3,531,600
450,000	0.772	0.3523	8,105	4,930	0.414	1.205	......	......	1.181	......	......	1.687	......	3,170,700
400,000	0.725	0.3143	7,230	4,400	0.368	1.130	......	......	1.150	......	......	1.612	......	2,828,700
350,000	0.679	0.2751	6,330	3,850	0.322	1.055	......	......	1.119	......	......	1.538	......	2,475,900
300,000	0.621	0.2359	5,425	3,300	0.276	0.973	......	......	1.081	......	......	1.454	......	2,123,100
250,000	0.567	0.1964	4,515	2,750	0.230	0.894	......	......	1.045	......	......	1.375	......	1,767,600
0000	0.522	0.1661	3,870	2,330	0.195	0.831	......	......	1.019	......	......	1.312	......	1,494,900
000	0.464	0.1319	3,165	1,850	0.155	0.755	......	......	0.976	......	......	1.234	......	1,187,100
00	0.414	0.1043	2,505	1,460	0.123	0.691	......	......	0.943	......	......	1.168	......	938,700
0	0.368	0.0830	1,990	1,160	0.097	0.637	......	......	0.912	......	......	1.112	......	747,000
1	0.328	0.0660	1,585	925	0.077	0.592	......	......	0.885	......	......	1.065	......	594,000
2	0.291	0.0520	1,250	730	0.061	0.553	......	......	0.861	......	......	1.023	......	468,000
3	0.261	0.0413	990	580	0.049	0.522	......	......	0.841	......	......	0.990	......	371,700
4	0.231	0.0328	790	460	0.039	0.494	......	......	0.821	......	......	0.958	......	295,200

NOTE.—A Loading = dead load and 15.0 lb. per square foot wind pressure.
B Loading = dead + 0.5-in. ice and 8.0-lb. wind pressure.
C Loading = dead + 0.75-in. ice and 11.0-lb. wind pressure.

TABLE 11.—PROPERTIES OF GALVANIZED STRANDED STEEL CABLE

Gage	Area, sq. in.	Guy wire, 60,000	Siemens-Martin, 75,000	High strength, 150,000	Ex. high, 180,000	Dead	Dead + 0.5-in. ice	Dead + 0.75-in. ice	15.0 lb. per sq. ft.	8.0-lb. + 0.5-in. ice	11.0-lb. + 0.75-in. ice	Load A	Load B	Load C	E = 25,000,000	E = 29,000,000
7/8	......	......	......	......	......	......	......	......	......	......	......	......	......	......	......	......
3/4	0.2356	14,100	19,000	35,300	42,400	0.815	1.515	......	......	1.083	......	......	1.862	......	5,890,000	6,832,400
5/8	0.1922	11,500	14,500	28,800	34,600	0.668	1.329	......	......	1.042	......	......	1.689	......	4,805,000	5,573,800
9/16	......	......	......	......	......	......	......	......	......	......	......	......	......	......	......	......
1/2	0.1443	8,500	11,000	21,600	26,000	0.510	1.132	......	......	1.000	......	......	1.510	......	3,607,500	4,184,700
7/16	0.1204	6,500	9,000	15,000	22,500	0.415	0.998	......	......	0.958	......	......	1.383	......	3,010,000	3,491,600
3/8	0.0832	5,000	6,800	12,500	15,000	0.295	0.839	......	......	0.917	......	......	1.243	......	2,080,000	2,412,800
5/16	0.0606	3,800	4,860	9,100	10,900	0.210	0.715	......	......	0.875	......	......	1.130	......	1,515,000	1,757,400

NOTE.—A Loading = dead load and 15.0 lb. per square foot wind pressure.
B Loading = dead + 0.5-in. ice and 8.0-lb. wind pressure.
C Loading = dead + 0.75-in. ice and 11.0-lb. wind pressure.

TABLE 12.—PROPERTIES OF TELEPHONE WIRES

Gage	Diam.	Area	Galv. BB 60,000	Galv. EBB 75,000	Hard copper, 60,000	Steel river crossing wire, 187,000	Dead (copper)	Dead + 0.5-in. ice	Dead + 0.75-in. ice	8.0-lb. + 0.5-in. ice	Load B	E = 16,000,000 copper	E = 29,000,000 steel
6 B.W.G.	0.203	0.0324	1,950	1,770	......	6,060	0.082	......	......	......	......	......	939,600
8 B.W.G.	0.165	0.0214	1,290	1,170	1,280	4,000	0.039	0.496	0.936	0.777	0.922	342,400	......
9 B.W.G.	0.148	0.0103	1,040	940	620	1,930	0.037	0.421	0.845	0.743	0.854	164,800	......
	0.110	0.0095	......	......	570	1,780	0.033	0.416	0.802	0.740	0.849	152,000	......
12 B.W.G.	0.109	0.0093	560	510	......	1,740	......	......	......	......	......	148,800	......

NOTE.—A Loading = dead load and 15.0 lb. per square foot wind pressure.
B Loading = dead + 0.5-in. ice and 8.0-lb. wind pressure.
C Loading = dead + 0.75-in. ice and 11.0-lb. wind pressure.

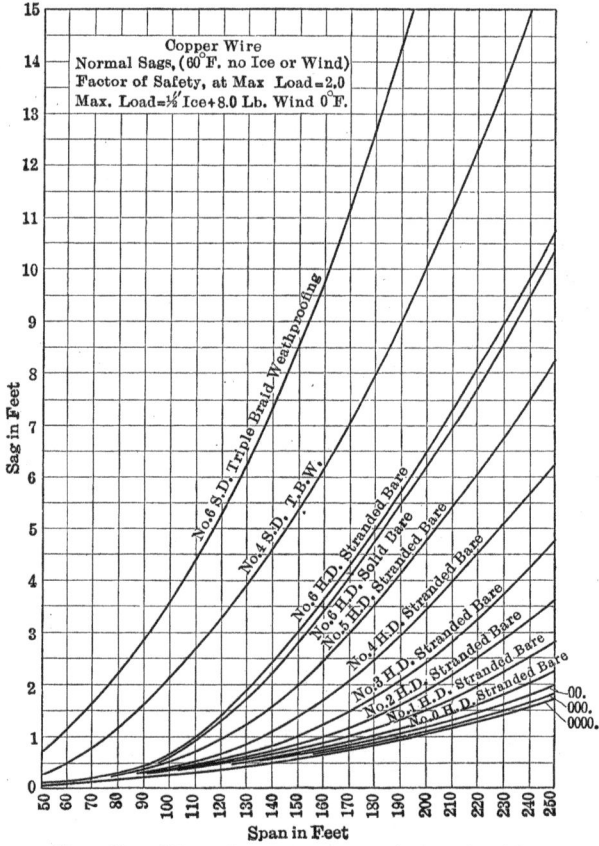

Copper Wire
Normal Sags, (60°F. no Ice or Wind)
Factor of Safety, at Max Load=2,0
Max. Load=⅛″ Ice+8.0 Lb. Wind 0°F.

FIG. 27.—Normal sags,[1] copper wires and cables.

[1]Overhead Line Construction Committee (N.E.L.A., 1911).

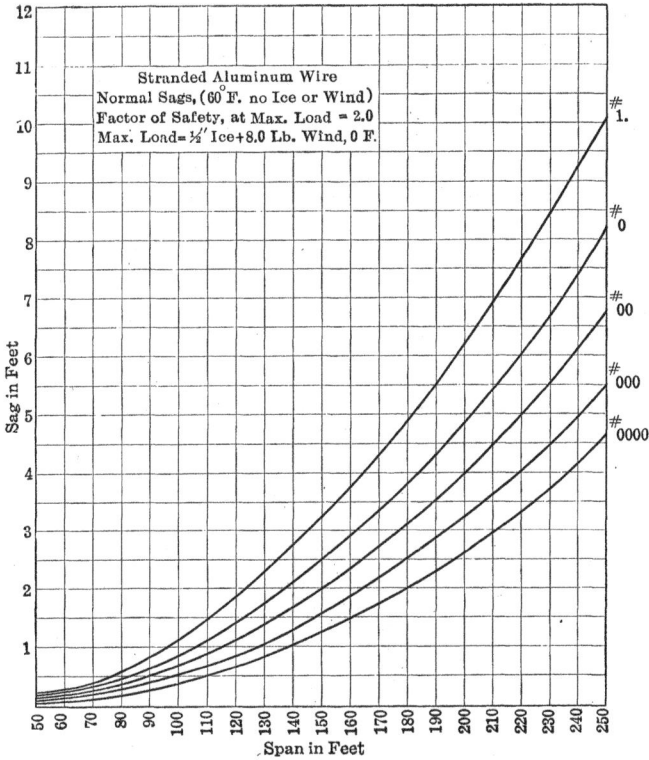

FIG. 28.—Normal sags,[1] aluminum cable

TABLE 13.—PROPERTIES OF WIRE MATERIAL

(From 1911 Report of Overhead Line Construction Committee of National Electric Light Association)

	Ultimate strength per sq. in.	Elastic limit	Modulus elasticity, E	Coefficient of expansion
Copper, solid, soft-drawn......	32–34,000	28,000	12,000,000	0.0000096
Copper, solid, hard-drawn.....	50–55–57–60,000	30–32–34–35,000	16,000,000	0.0000096
Copper, stranded, soft-drawn..	34,000	28,000	12,000,000	0.0000096
Copper, stranded, hard-drawn.	60,000	35,000	16,000,000	0.0000096
Aluminum, stranded..........	23–24,000	14,000	9,000,000	0.0000128
Steel, stranded, Siemens-Martin	75,000	....	29,000,000	0.0000064
Steel, stranded, high-tension...	125,000	....	29,000,000	0.0000064
Steel, stranded, ex-high-tension	187,000	....	29,000,000	0.0000064

[1] Overhead Line Construction Committee (N.E.L.A., 1911).

CHAPTER IV

DESIGN

Since it is impracticable to include herein a sufficient explanation of the laws of mechanics or the theory and practice of structural design, to enable the inexperienced to acquire even a reasonable facility in their use, no detailed exposition thereof has been attempted. The computation of stresses, and the determination of sections for such structures as steel towers, requires a working knowledge of subjects already covered by various text-books. However, there are a number of general conditions in which transmission-line work does not follow the

Fig. 29.—Relative strength of telephone and power lines.

accepted standards and methods of other structural design, and a discussion of such matters should be of value to otherwise competent designers whose experience has been obtained in a different field.

Factors of Safety, Etc.—If a given line is to be designed in a logical manner and with a minimum of "cut and try" methods, the first step is the assumption of the various loads and factors of safety. These assumptions will enable the designer to mentally predetermine, to some extent, the general nature of the supports, or at least to narrow the field of choice. In a broader sense, it will also limit the choice to one kind of material for the supports, since a wooden-pole line cannot have a total factor of safety equal to that possible in steel construction.

The first and easiest factor of safety to assume is that for the wires. It is the easiest to assume, since once chosen it can be maintained without much effect upon the type of support. Again, a reasonable factor in the wires will have only a beneficial effect upon all the remaining construction. Further, there is a more general consensus of opinion regarding this assumption than on any other element affecting transmission-line construction. The literal expression of various wire loads and factors may appear quite dissimilar, but the ultimate result when the wires are strung presents a fair average. Unbalanced loads and factors of safety have been very common, and they are greatly to be condemned both as a misstatement of fact and as providing an excuse for future errors of design.

Sleet may be encountered, during the probable life of a well-built transmission line, in a great many sections throughout America, even in the South. Moreover, it may be desirable to provide a sag corresponding to the standard sleet load, even in non-sleet regions, because such sags decrease both the normal and the maximum loads on the line and, in general, produce a line which is well able to distribute and equalize excess stresses.

In general the "½-in. ice plus 8.0-lb. wind" loading is logical in origin, as shown by the writer in 1908;[1] further it is more universally accepted than any other.

A factor of safety in wires of 2.0 is approximately equivalent to a working stress of 0.9 of the elastic limit, and so is not too conservative when errors of stringing and possible reduction of strength at splices are considered.

An exception may be made to the above loads and factor in the case of very long spans, in which the load may be consistently reduced about 25 per cent. Thus in designing one line with 800-ft. spans, the writer believes that his use of a 6.0-lb. wind pressure was logical. An additional exception to the above factor, if not to the load, is the short-span distribution lines in city streets. Such lines are usually well sheltered, designed for low voltages, heavy and numerous wires, and better guyed

[1] Proceedings American Society of Civil Engineers, 1908.

when the lateral restraint of wires and the guys at street corners are considered. As proven by actual experience with many thousands of miles of such lines, they may consistently be given a factor of one (1) with the above maximum loading.

It is true that many transmission lines have been built with tighter stringing than recommended above—only a few having had more conservative values—but as a general statement, slack lines are safer. If the number of wires will permit liberal separation of the conductors, a little increase in the sag will often help out in the design of the supports. It seems probable that for every span in which slack stringing (and improper separation) have caused accidental contacts there have been three cases in which tight stringing has broken pins, insulators or supports.

After making the foregoing assumptions there still remain to be determined certain loads and factors for the supporting structures. The specified wire load applies also to the supports, but in addition there is the broken-wire load to be considered, as well as the factor, or factors, for the poles or towers. Broken-wire loads are discussed in more detail elsewhere (pages 39 to 42), the writer's recommendation being that for transmission lines the effect of one broken wire should be combined with the above-mentioned ice and wind loads, while for short-span city distribution lines broken-wire loads may be neglected.

While the requirements just mentioned would literally apply to insulators and pins on all supports, it is impracticable and unnecessary to so interpret the broken-wire load. It will, therefore, apply only to insulators and pins at corners, dead ends, crossings and special points. On intermediate tangent poles single-pin insulators and tie wires may properly be used, although they may not always have the required broken-wire strength, and the broken-wire effect on the pole might be considered as the equivalent result of one broken wire or of several unbalanced wires.

Owing perhaps to unfamiliarity with the structural questions involved in the construction of transmission lines, engineers have, until recently, specified the test loads which sample towers or poles must withstand. Dependence on a large, and more or less certain, factor of safety to cover uncertain design, however, should have no permanent place in line construction. On the other hand, the theory of this practice, *i.e.*, that of working to known ultimate strengths, has much to commend it. Moreover, tests of sample towers have been of considerable use in adding to our somewhat meager store of data on the ultimate strength of columns.

Unfortunately for the entire success of this procedure, a test load is very rarely an accurate representation of the maximum which may be obtained in practice, nor is the condition of the test structure similar to that of many of the structures when installed. Test loads are almost always applied regularly and slowly, and in many cases uneccentrically. A test structure most assuredly will have at least a fairly good foundation, and be composed of members free from incipient bends or other defects caused by mishandling. It would also be very well bolted together and plumbed with greater accuracy than the average line structure. In general it may be said that an expert structural assembler should be able to obtain test loads quite noticeably in excess of the presumptive average strength of the finally erected structures. It appears, therefore, that the period of usefulness for this practice is past, and that competent designers should be able to produce structures which will have actual strengths much nearer their predetermined strengths than the actual loads will be to the assumed loads. If statically indeterminate frames are used, such as poles with incomplete web systems, design tests are necessary, but the transfer of stresses and the construction of efficient details are now so well understood that actual tests of determinate structures are in a measure a confession of ignorance.

The failure of steel poles and towers has almost invariably been caused by the buckling of main compression members, and this may or may not be superinduced by inefficient bracing. Owing to the possible application of the load from the opposite side of a structure, line supports must have the same main compression section at each corner regardless of the tension stress. The compression stress per square inch in the main legs is, therefore, the first and most important determination. A secondary condition which should be borne in mind during the calculations is that the section selected must be of a size suitable for the connection of the desired bracing.

Admitting that there is a difference in the kind of service or, in all events, in the number of applications of the load, between transmission and building work, there is a very marked variation in the attitude of engineers toward bolted connections in the two types of construction. In building work single bolts are discouraged or, if used, low strength values are allowed. In towers, however, all joints are bolted, and then usually with one-bolt connections in which no reduction of value is assumed.

The theoretical value of a one-flange connection should be reduced as the full strength of the connected member is not available; this has not been so assumed in tower work. Again, the bearing of a bolt is assumed as being on a surface as thick as the member; in fact the bearing will be only on a line.

The strength of a one-flange connection is approximately but 80 per cent. of the strength of the angle; therefore, since many tower joints are one-flange connections, a somewhat conservative unit stress should be assumed in the design of bracing. Further, if a member has one flange "blocked off," *i.e.*, cut entirely away for clearance, or if the flanges are mashed together or flattened out, the strength of the member at that point is no longer the strength of an angle but of a flat. In addition, there is considerable likelihood that such blacksmith work may result in burning or cracking the material at the point in question.

As a long slender member is not well adapted to withstand compression, it has been customary in other work to limit the relation of the length to the radius of gyration. In transmission-line construction, however, very much higher values of this ratio have been used than are generally permitted. It is probably not necessary to adhere to the low limits of building construction, but it is equally probable that too much latitude has been taken in some cases heretofore.

The more recent designs of transmission line supports do not employ any castings in the main structure, although cast-steel or malleable-iron castings are perhaps properly applicable to wire connections. Similar reasoning should prohibit the use of castings for hoops or bands in reinforced-concrete poles.

Transverse Loads.—Before entering upon any detailed discussion of design, it is necessary to consider briefly the forces acting upon a pole line and the character of service required of its component parts. As already stated, the function of the pole is that of a cantilever beam rather than of a column. The external forces are due to dead, ice, and wind loads, which, with the exception of the pressure on the pole, must be transmitted to the pole by the wires.

The weight of the wires and their coating of sleet, together with the weight of crossarms, insulators, and the pole itself, is a vertical load which the pole carries as a column. The pressure of the wind on the wires, whose diameter is increased by the sleet, and upon the pole structure, is assumed as acting horizontally and at a right angle with the line, and,

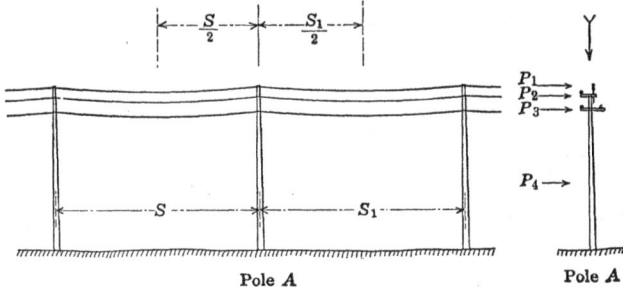

Fig. 30.—Transverse loading.

P_1 = Transverse load at ground wire = wind load per ft. of wire $a\left(\dfrac{S}{2}+\dfrac{S_1}{2}\right)$.

P_2 = Transverse load at top power wire = wind load per ft. of wire $b\left(\dfrac{S}{2}+\dfrac{S_1}{2}\right)$.

P_3 = Transverse load at lower power wires = wind load per ft. of wire $b\left(\dfrac{S}{2}+\dfrac{S_1}{2}\right)\times 2$.

P_4 = Wind load per ft. of pole $\times$ length of pole above ground.

V = Vertical load on pole = weight per lin. ft. of wire $\times$ no. of wires $\times \left(\dfrac{S}{2}+\dfrac{S_1}{2}\right)$
$\qquad$ + weight of pole.

therefore, its effect is much greater than the effect of the vertical forces.

When poles are closely spaced there is undoubtedly some side-guying effect due to lateral restraint from the wires. That is, if one pole is subjected to a severe gust of wind, the neighboring poles will be brought into action, to a limited extent, by the wires spreading out some of the load to adjacent poles.

Corner Loads.—In the case of a pole placed at a bend in the line, there must be *added* to the foregoing the transverse component of the tension in the wires, *i.e.*, the maximum tension multiplied by twice the sine of one-half the angle of the bend.

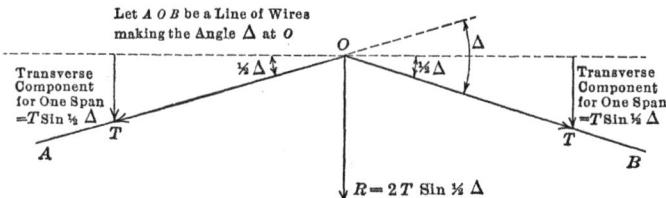

Fig. 31.—Corner loading.

A little study of Fig. 32 will explain why corners are usually the weakest points in a line, since they are unquestionably and invariably the only supports subject to continual loading and because *a portion of the load alone may be considerably more than a dead-ending load.* In the writer's opinion, considerably more effort and money than has usually been expended is amply warranted in avoiding or decreasing curves and corners.

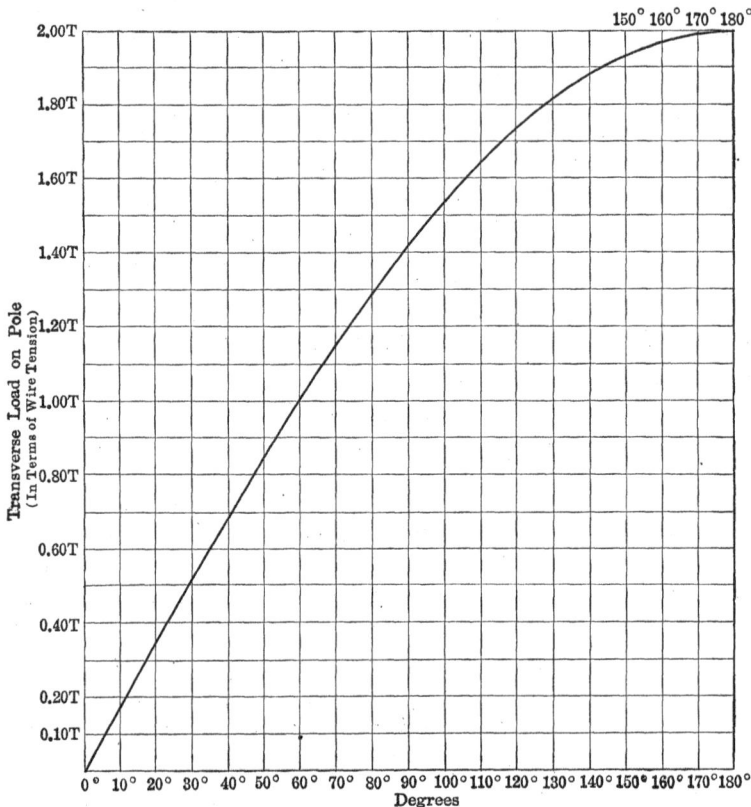

FIG. 32.—Corner pole loading.[1]

Following the common practice in low-voltage construction, the statement is frequently made that at corners the neighboring spans should be shortened to minimize the stress on the corner structure.

In the absence of further elaboration, and in view of the usual lack of expert advice during erection, it would appear that the above statement is thought to be self-sufficient.

[1] R. D. Coombs & Co. Design Standards.

This is decidedly not the case, since the stress to which the corner support is subjected is not reduced materially by shorter spans, unless advantage is taken of the short spans to increase the sag in those spans. This is due to the fact that the greater part of the load upon the corner support is from the tension in the wires, and unless this tension is reduced by slack stringing, there is no particular advantage in short spans.

With short-span low-voltage construction, an increase in sag of a few inches, made by the line foreman to "ease up the corner," will be inconspicuous and very efficient, provided that the tying-in of the wires is effective for some distance each side of the corner. Long-span high-voltage construction, however, requires a material change in sag and there is a redistribution of stresses to be provided for, unless entire dependence is to be placed on the bending of supports and on the slipping of wires at the supports.

Slack corner spans may have a maximum wire tension from 500 to 1000 lb. less than the standard stringing; so if this reduction is to remain effective the unbalanced tension must either be held by and at the adjoining supports or be carried back and distributed over a number of supports.

It is needless to say that the standard stringing curve, in case one is provided, is not applicable to such construction. Further, it is useless to attempt to distribute unbalanced wire tension by means of a slip-shod single pigtail tie wire.

Broken-wire Loads.—In case the sags in adjoining spans are not adjusted so as to balance the tension in the wires each side of a pole, there will be an unbalanced pull in the direction of the line, which must be considered in conjunction with the vertical and horizontal forces first mentioned. Unbalanced tension may also be produced by unequal ice and wind loads in adjoining spans. Further, if it is assumed that all, or part, of the wires may be broken, then the poles must withstand a longitudinal force equal to the tension in the wires in the unbroken span.

On the other hand, it can be shown by a rather complicated mathematical demonstration that, owing to certain properties of the catenary curve, a slight bending in a number of poles will balance the tensions in adjoining spans. This is due to the fact that the tension in a wire is greatly decreased if the span length is shortened while the length of wire per span remains unchanged. Vice versa, increasing the span length while the length of wire per span remains unchanged increases the tension.

If it is assumed that all the wires in one span are broken, then the first pole is subject to the unbalanced tension of all the wires in the unbroken span and bends away from the break. This shortens the next span length, decreases the tension in that span, and allows the second pole to be bent away from the break. Successive bending occurs in decreasing amounts, until a point of equilibrium is reached at which the wire tension next to the break is considerably less than the original tension.

If it is assumed that less than the entire number of wires are broken, then the bending of the first pole increases the span length of the remaining wires and by increasing their tension causes them to exert a greater pull toward the break and thus decreases the unbalanced pull on the pole.

However, the ordinary attachments for fastening line wires to the insulators do not always have sufficient strength to develop the strength of the wire and, therefore, a broken wire would pull through into the adjoining spans before exerting its maximum tension on the poles. For this reason, and because equilibrium by bending may result in over-stressing the poles, wires, pins or insulators, it is not always possible to take advantage of this method of design.

In assuming the possibility of broken wires, it becomes necessary to assume which wires may break as well as their number. If the wires farthest from the pole are broken, the effect on the crossarms is much greater than in the case of wires near the pole. If all the broken wires are on one side of the pole, the torsional effect on the pole must be considered.

Column Formulas.—Inasmuch as the strength of the main-leg members of a pole or tower, as well as most of the bracing, is predicated upon their strength as compression members, the most important requirement of a specification next to the broken-wire condition is the formula for compression members, known as the column formula. Unfortunately, the many column formulas in existence are always expressed in terms of "safe working unit stresses," which renders them almost valueless to the inexpert transmission-line designer, unless their factor of safety is known. This is due to the fact that in transmission line construction it is the ultimate or breaking strength which must be determined in order that a specified factor of safety may be applied thereto.

The writer is aware that the last statement may be criticised as a defense of a dangerous practice, in that no protection is afforded by the working stress for incompetence on the part of the designer. On this point the writer wishes to emphasize his belief that, provided a portion of the factor of safety is present to offset minor errors of design, the unit stress should not be expected to afford such protection. As a matter of fact, when bidding under a specification requiring a test in which a sample pole or tower "must withstand" certain loads, the competing designers are compelled to work as close to the probable ultimate or buckling strength as seems to them advisable. Therefore, it would serve to eliminate the personal equation of the designer, together with the false security arising from test towers, if there were available accurate rules by which to compute the ultimate strength of statically determinate structures.

In general there are such rules, and aside from the difficulties or inaccuracies of computing eccentric or torsional stresses, the chief uncertainty is in the column formula. The engineering profession has long awaited a complete series of ultimate compressive tests and the derivation therefrom of a set of generally accepted column formulas.

It is to be regretted that the hundreds of tower tests which have been made to date have not resulted in a more accurate and more general addition to our knowledge of the subject.

In pole and tower design, the compression members are simple in type, usually single angles with relatively large ratios of the unsupported length to the radius of gyration, *i.e.*, $\dfrac{l}{r}$. "Failure" occurs when such members buckle, as the structure becomes distorted and useless, although it may not fall to the ground. It is readily apparent that any incipient bends in such columns will very markedly affect the theoretical compressive strength. In addition it is quite possible to select sections such as 4 in. $\times$ 4 in. $\times \frac{1}{4}$-in. angles, for example, whose theoretical strength by the column formula exceeds their actual strength. This is due to the fact that in such large thin sections failure may start by the local buckling of the legs of the angle.

Columns have been divided into classes according to the nature of their end connections, whether "fixed," "pin-ended," or "round-ended" and free to move. A pole set in an adequate concrete foundation probably approaches the condition of one fixed and one free end, if the whole pole is under consideration. The columns formed by tower members in general might be considered as stronger than flat-ended and weaker than fixed-ended columns.

A discussion or compilation of the various column formulas and their derivation is beyond the scope of this book, but in Fig. 31 are shown several formulas expressed in terms of ultimate

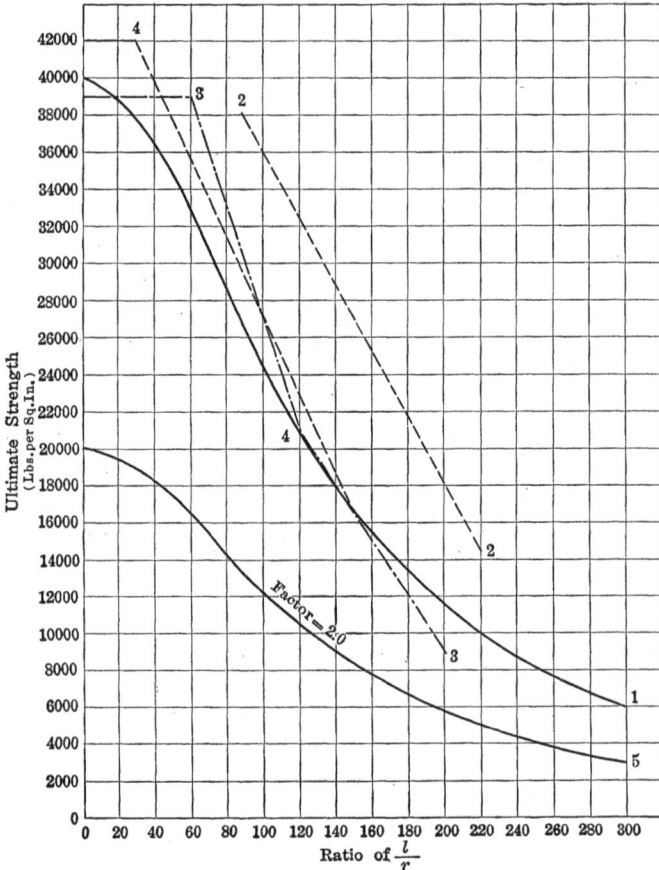

FIG. 33.—Column formulas.

strength. It will be noted that the chief differences are at the ends of the curves, *i.e.*, either for very small or very large values of *l/r*. *It should be noted that the ultimate-strength curves are based on the assumption of good design, workmanship and material, the use of medium steel, and freedom from incipient injury. They represent approximate average values and must be used with caution by the inexpert.* They are chiefly useful in showing the derivation of the allowable working values, and to predetermine breaking strengths.

Ultimate-strength Curves:

(1) R. D. Coombs & Co.: $\dfrac{40000}{1 + \dfrac{1}{16,000}\dfrac{l^2}{r^2}}$

(2) Joint Report Crossing Specifications: $54,000 - 180\,\dfrac{l}{r}$

(3) American Bridge Co.: $\begin{cases} 57,000 - 300\dfrac{l}{r} \\ 39,000 - 150\dfrac{l}{r} \\ 39,000 \text{ max.} \end{cases}$

(4) American Railway Engineering Association: $\begin{cases} 48,000 - 210\dfrac{l}{r} \\ 42,000 \text{ max.} \end{cases}$

Working-stress Curve:

(5) R. D. Coombs & Co. (with a factor of safety of 2.0): $\dfrac{20,000}{1 + \dfrac{1}{16,000}\dfrac{l^2}{r^2}}$

or

(with a factor of safety of 2.5): $\dfrac{16,000}{1 + \dfrac{1}{16,000}\dfrac{l^2}{r^2}}$

Formulas 2, 3 and 4 shown in Fig. 31 were not issued in that form by their authors, but in terms of allowable unit stresses, as follows:

2. $18,000 - 60\,\dfrac{l}{r}$

3. $\begin{cases} 19,000 - 100\,\dfrac{l}{r} \\ 13,000 - 50\,\dfrac{l}{r} \\ 13,000 \text{ max.} \end{cases}$

4. $\begin{cases} 16,000 - 70\,\dfrac{l}{r} \\ 14,000 \text{ max.} \end{cases}$

The diagrams shown were, therefore, obtained by increasing these allowable units to their apparent ultimate values. The difficulty in comparing the usual allowable-unit formulas arises from their variable and uncertain factors of safety. Such formulas were rarely intended for use with the large ratios of l/r com-

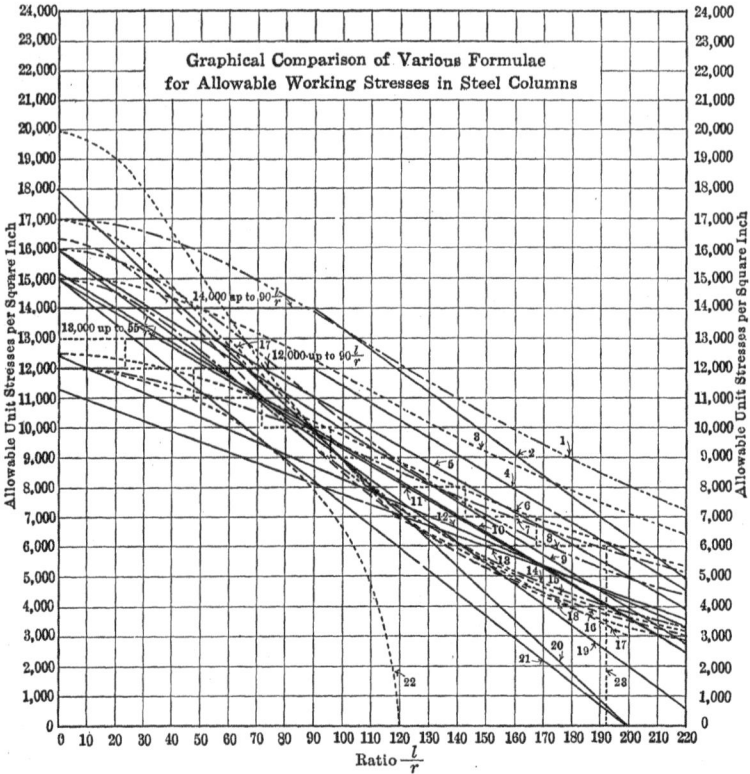

FIG. 34.

1. $\dfrac{17,000}{1+\dfrac{l^2}{36,000r^2}}$ Osborne (Ry. Specificat'ns)

2. $20,300-70\dfrac{l}{r}$ Chicago Bridge & Iron Wks. (Tank Towers)

3. $\dfrac{15,000}{1+\dfrac{l^2}{36,000r^2}}$ Chicago, Milwaukee

4. $17,100-57\dfrac{l}{r}$ Carnegie, Jones & Laughlin, Buffalo-Minneapolis

5. $16,000-55\dfrac{l}{r}$ Bethlehem Steel Co.

6. $\dfrac{12,500}{1+\dfrac{l^2}{36,000r^2}}$ Cambria, Carnegie, Jones & Laughlin, Phoenix Bridge Co.

7. $\dfrac{12,000}{1+\dfrac{l^2}{36,000r^2}}$ Boston

8. $\dfrac{16,000}{1+\dfrac{l^2}{18,000r^2}}$ Dominion Government

9. $16,000-60\dfrac{l}{r}$ J. A. L. Waddell ("De Pontibus")

10. $15,200-58\dfrac{l}{r}$ New York City, Washington Nat'l. B'd. Fire Underwriters

11. $15,000-57\dfrac{l}{r}$ Passaic Steel Co.

12. $12,500-42\dfrac{l}{r}$ C. E. Fowler ("Steel Roofs & B'ldgs.")

13. $11,300-35\dfrac{l}{r}$ H. E. Horton

14. $\dfrac{15,000}{1+\dfrac{l^2}{13,500r^2}}$ Chesapeake & Ohio Ry., Norfolk & Western Ry., Phila. & Reading Ry., Penna. R. R., N. Y. C. (for H'way Bridges)

15. $\dfrac{17,000}{1+\dfrac{l^2}{11,000r^2}}$ Virginia Br. & Iron Co., Baltimore & Ohio Ry., Chesapeake & Ohio Ry., Long Island R. R., Deepwater Ry., Phila. & Reading Ry.

16. $\dfrac{16,250}{1+\dfrac{l^2}{11,000r^2}}$ Philadelphia

mon in tower work, nor were the conditions of loading with which they were used as unusual as those encountered in transmission-line construction. In other words, tower members are less likely to approach their theoretically perfect condition and, therefore, should have a high factor of safety, while the loading for which they are usually figured may never occur, and on that account they should have a low factor.

TABLE 14.—ULTIMATE STRENGTHS OF TIMBER IN BENDING

	N.E.L.A. Overhead Line Construction Committee, 1911	A.R.E.A. Wooden Bridges and Trestles Committee, 1909
	Lb. per square inch	Lb. per square inch
Port Orford cedar	6900	
Long-leaf yellow pine	6600	6500
Douglas fir	6000	6100
Short-leaf yellow pine	5700	5600
White oak	5700	5700
Chestnut	5100	
Washington cedar	5100	
Idaho cedar	5100	
Redwood	5100	5000
Bald cypress (heartwood)	4800	4800
Red cedar	4200	4200
Eastern white cedar	3600	
Juniper	3300	
Catalpa	3000	
Spruce		4800
Western hemlock		5800

It should be noted that the extension of the ultimate-strength curve, 1, to values of l/r of 300 is not intended as a recommendation of such values, but rather to illustrate the decrease of strength. Further, even the reduced values shown will not always be obtained in practice owing to errors of design, workmanship, and injuries from handling.

17. $19,000 - 100 \frac{l}{r} \left(\frac{l}{r} \text{ to } 120\right)$ Amer. Bridge Co.

$13,000 - 50 \frac{l}{r} \left(\frac{l}{r} 120 \text{ to } 200\right)$

13,000 Maximum

18. $\dfrac{20,000}{1 + \dfrac{l^2}{8,000r^2}}$ H. B. Seaman

19. $16,000 - 70 \frac{l}{r}$ Am. Ry., Eng. & M. W. Assn., N. Y. C. R. R. Bos. & Me. Ry., Canadian Pacific Ry., Grand Trunk Ry., Maine Central Ry., Mo., Kansas & Tex. Ry., Nash.,

Chatt. & St. L. Ry., N. Y., N. H. & H. Ry. St. Louis & San Fran. Ry., St. Louis & Southwestern Ry., Wabash Ry., M. S. Ketchum ("Steel Mill Bldgs."); C. C. Schneider ("Structural Design of Bldgs.")

20. $18,000 - 90 \frac{l}{r}$ Southern Ry.

21. $15,000 - 75 \frac{l}{r}$ L. D. Rights

22. $12,000 \sqrt{1 - \dfrac{l^2}{(120r)^2}}$ J. R. Worcester

23. Rule. J. R. Worcester

The logical procedure would seem to be to assume only the loads which could reasonably be expected, *i.e.*, the ice and wind loads with very moderate broken-wire conditions, to prohibit excessive values of l/r, as well as the very thin sections, and to use a fairly low factor of safety.

The American Railway Engineering Association values are for squared timbers used in railroad construction, assumed to be of good commercial quality, and neither very green nor seasoned. It should be remembered that these values are the breaking strengths of fairly good lumber, and a more conservative factor of safety must be assumed than would be necessary in more uniform material. Further, pole timber is not squared and is more likely to contain crooked grain, knots and rot. It would be more consistent to use lower ultimate strengths for poles than the N.E.L.A. values, and a smaller factor of safety than 6.0, which is that used therewith.

Strength of Wooden Poles.

WEAKEST POINT IN WOODEN POLES

(a) *Neglecting Wind on Pole*

y = distance of weakest section below load.
d = diameter of pole at load.
t = increase in diameter per inch of length.
P = resultant load.

All dimensions in inches.

$$S = \frac{M}{\dfrac{I}{c}} = \frac{Py}{0.098(d + ty)^3}$$

$$\frac{ds}{dy} = \frac{0.098(d + ty)^3 P - 3 \times 0.098 P y t (d + ty)^2}{0.098^2 (d + ty)^6} = 0$$

$$(d + ty) - 3ty = 0$$

$$y = \frac{d}{2t}$$

Let d_1 = diameter of pole at distance y below the load. Then

$$t = \frac{d_1 - d}{y}$$

From above equation,

$$y = \frac{d}{2t} = \frac{dy}{2(d_1 - d)}$$

$$2(d_1 - d)y = dy$$

$$2d_1 = 3d$$

$$d_1 = \frac{3d}{2}$$

That is, the diameter of the pole at the weakest section is one and one-half times that at the load. When the diameter at the ground line is less than one and one-half times the diameter at the load, the weakest section is at the ground line.

WEAKEST POINT IN WOODEN POLES

(b) *Including Wind on Pole*

y = distance below load P.

d = diameter of pole at load.

t = increase in diameter per inch of length.

a = distance of load P below top of pole.

w = wind load on pole per inch of length.

$$S = \frac{M}{\dfrac{I}{c}} = \frac{Py}{0.098(d + ty)^3} + \frac{w(y + a)^2}{2 \times 0.098(d + ty)^3}$$

$$y = \frac{2P}{w} - \frac{d}{t} + 2a \pm \sqrt{\frac{4P^2}{w^2} - \frac{2Pd}{wt} + \frac{8Pa}{w} + \frac{d^2}{t^2} - \frac{2da}{t} + a^2}$$

or

$$y = \frac{2P}{w} - \frac{d}{t} + 2a \pm \sqrt{\left(\frac{2P}{w} - \frac{d}{t} + 2a\right)^2 + \frac{2d}{t}\left(\frac{P}{w} + a\right) - 3a^2}$$

CHAPTER V

WOODEN POLES

The total number of wooden poles in use in the United States is probably 40,000,000, while the yearly additions approximate 4,000,000. Of these the majority are used by telephone, telegraph, and railroad companies. The greater part of the 4,000,000 new poles are less than 40 ft. long, and such poles

FIG. 35.—Wooden pole lines, 60,000 volts.

are rarely used for transmission lines. The lengths of the poles used for transmission lines are increasing, while those for other purposes are decreasing. In the East the timber generally used is chestnut, while cedar is more common in the West. For distribution lines not only the length of the poles is increasing, but also the number installed per year. The cost of poles increases very rapidly with increases in length.

Approximately 90 per cent. of the timber poles used are either chestnut or cedar, the former being about 18 per cent. and the latter 72 per cent.

Chestnut, which is second in use to cedar, is durable and is stronger than cedar, and its taper is not excessive. On the other hand, it is heavier and harder, it shrinks and checks more easily, and is not so straight or free from knots.

Cypress is durable, but its size and taper often make it unsuitable for pole purposes. Yellow pine is not durable and is heavy, but it grows in suitable sizes.

Douglas fir, spruce and redwood are durable and make excellent poles when of suitable size. The consumption of the

FIG. 36.—Ring shakes in chestnut.

FIG. 37.—Cat-faces in chestnut.

former for pole purposes is increasing. Redwood, on account of its size, is used almost exclusively in the form of sawed poles.

The useful life of a timber pole, in contact with the soil, depends in part on the chemical action of the earth's ingredients, the attack by fungi, and on the ability of the timber to resist insects. Disintegration will, therefore, advance more rapidly in some soils than in others, but in general the use of good native timber

for local use will be found advisable. Decay at the ground line
weakens the body of a pole until this critical section is so emaciated
that it will no longer sustain its load. In the dry season this
decayed portion is much in the nature of dry tinder, so if the
pole is located on a grassy right-of-way, grass fires may char
away still more of the critical section.

Decay and Defects.—The decay of wood is generally due to
the activities of certain low forms of plant life known as fungi,
punk, toadstools, etc. Bacteria are also known to cause decay,
but their action is not well understood. These plants have their
origin in minute spores borne from place to place by the wind.

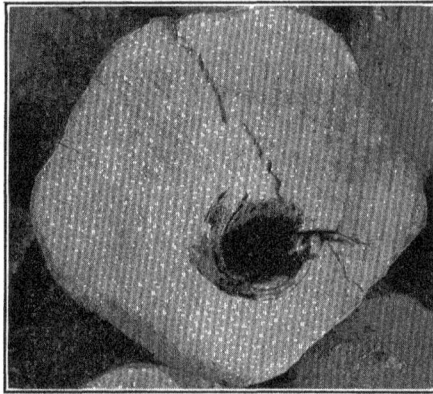

Fig. 38.—Butt-rot in chestnut.

Those that lodge in a suitable situation for growth, which may be
on either living or dead timber, germinate, and provided the
conditions are favorable, at once attack the wood. The plants
grow with great rapidity, sending out numerous threads which
penetrate the wood and attack the contents of the wood cells and
finally the cell-walls.

The most favorable conditions for the growth of fungi and
other organisms of decay are an abundant food supply, heat,
moisture, and air, the amount of each depending on the kind of
organism. A certain amount of moisture must be present or decay cannot set in. Air is also essential, and thus may be explained
the lasting qualities of wood when kept perfectly dry, and the
perfect state of preservation of wood which has been under water
for long periods, moisture being lacking in the first case and air
in the second. Again, if the wood is protected by a germicide or

antiseptic, it will not decay. At the butt of the pole, though moisture is present, air is excluded, while above the ground the pole is generally dry. Decay begins where moisture and air are both present, the former perhaps being drawn by capillary attraction from the ground.

Since the decay of timber is due to the attacks of wood-destroying fungi, and since the most important condition of the growth of these fungi is water, anything which lessens the amount of water in wood aids in its preservation.

Cold, or extreme heat, will prevent the growth of fungi, although the necessary degree of the latter is beyond the limits

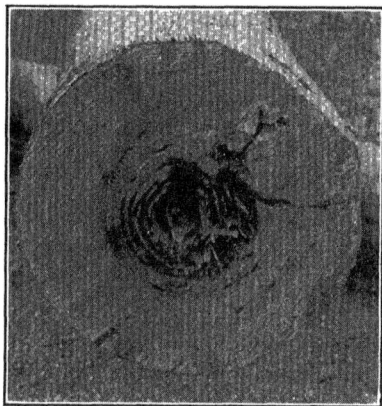

Fig. 39.—Ant-eaten butt.

of natural temperatures. The character of the soil may have a marked effect on the decay of timber, owing to the ability of certain soils, like heavy clay, to hold water or to discourage insect life.

The decay of poles before their installation in the line may be of several kinds and together with the several kinds of cracks or other injuries constitute what are known as "defects." The former include butt rot, heart rot, ring rot, and rotten knots; and the latter, seasoning checks, wind shakes or ring shakes, cat faces, and loose knots.

Since in designing the strength of a pole is computed as that of a solid homogeneous cylindrical section, it is evident that pronounced defects may materially affect the actual strength. Therefore, either pole specifications and inspection must be strict enough to eliminate poles whose actual strengths are not reasonably close to their assumed values, or else a large factor of safety must be employed to take care of irregularities.

Some detailed instructions to inspectors would seem very desirable, since exactly the same defect may have an entirely different significance in two locations on the pole. Thus a rotten heart, very common in cedar, may be in the center of the cross-section and, therefore, of the least effect, or it may be well toward one side. A wood pole is not greatly affected by hollowing out a small portion at the center of the cross-section, but the strength is decreased by any loss of area near the circumference or any reduction of diameter. The strength is proportional to the cube of the diameter, and the relative strengths for different conditions would be:

15-in. sound diameter	= 100%
15-in. pole, 5-in. rotten heart at center	= 99%
15-in. pole, 5-in. rotten heart 2½ in. off center	= 89%
12-in. sound diameter (1½ in. decay)	= 51%

Seasoning.—Under present methods much timber is rendered unfit for use by improper seasoning. When exposed to the sun and wind the water will evaporate more rapidly from the outer than from the inner parts of a log and more rapidly from the ends than from the sides. The evaporation of water from timber is largely through the ends. The evaporation from the other surfaces takes place very slowly out-of-doors, with greater rapidity in a kiln. The rate of evaporation differs with the kind of timber and its shape. Air-drying out-of-doors takes from two months to a year, the time depending on the kind of timber and the climate. As the water evaporates the wood shrinks, and when the shrinkage is not fairly uniform the wood cracks. When wet wood is piled in the sun, evaporation may occur with such unevenness that the timber splits and cracks so badly as to become absolutely useless. Such uneven drying can be largely prevented by careful piling. When solid piles are placed side by side and many together, the air cannot circulate freely between the timbers. Open-crib piles, however, will allow free air circulation even when closely spaced. For this reason green timber should be piled in as open piles as possible, as soon as it is cut, and kept so until it is air dry. No timber should be treated until it is air dry.

Seasoning is ordinarily understood to mean drying, but it

really involves other changes besides the evaporation of water. It is very probable that these consist in changes in the substances in the wood fiber, and possibly also in the tannins, resins and other incrusting substances.

One of the first steps in preparing naturally short-lived timber for preservative treatment is to season it properly. More benefit will result from taking care of the short-lived timbers than from treatment of those with longer life.

The bark should be peeled from poles before seasoning, and particularly from those that are to be treated, as the inner bark offers considerable resistance to impregnating fluids and if not removed will peel, leaving the untreated wood exposed to the attack of fungi. Bark will also retard and almost prevent seasoning. Care should be taken in handling and felling trees, as those which are split in felling or are otherwise roughly handled may afterward undergo serious checking. Whether poles are to receive preservative treatment or not, there can be no doubt that it invariably pays to season them properly before putting them into service. Under ordinary conditions, the life of a well-seasoned untreated pole should be at least 30 per cent. greater than that of an untreated green pole.

Poles should be cut from sound timber, which may or may not be live timber. Poles cut in the late winter or spring have immediately before them the best period for seasoning, but late fall and winter offer the best conditions for cutting, and facilitate the cultivation of new sprouts from the winter-cut stumps.

Preservatives.—The creosotes are employed most generally for protecting timber against decay, and they are apparently the best type of preservative. The terms creosote and tar are rather general expressions, and not very definitely interpreted by the ordinary purchaser. Briefly, coal-tar is produced by the distillation of coal and is obtained from two distinct and different sources, *i.e.*, that from coke ovens and that from gas works. Crude coal-tar is subjected to different refining processes, and yields various commercial derivatives, such as the different grades of creosote, carbolics, naphthas, etc. The protective effect of all preservatives is due to their exclusion of water and to their antiseptic or poisonous effect on the fungi which cause decay. In order to protect timber permanently, it is necessary that the preservative be maintained either as an impervious coating on the surface or as an impregnation through-

6

out at least the outer layers of the wood. A thin permanent surface coating is impracticable; therefore, to obtain successful results the preservative should be injected into the timber. The deeper the penetration and the more insoluble and non-evaporative the injected material, the more successful will be the treatment. It is possible to inject a light solution to greater depths than a heavy solution, but it may be that the lighter solution, which may contain more volatile matter, will evaporate or wash out more readily than a heavier solution. It is customary, however, to regard the greatest penetration as the most desirable, and to grade the treatment by the weight of preservative injected into a given volume of timber. Such units of measurement, however, are only comparable if the qualities of the injected fluids are similar.

In addition to the three regular grades of creosote oil, *i.e.*, the A.R.E.A. Nos. 1, 2 and 3, the specifications of which are recognized standards, there are in use various combinations of creosote and coal-tar as well as a number of proprietary creosotes, or creosote-tar combinations. Since creosote is obtained from tar, and is a part of crude tar, the distinction between creosote, as the word is ordinarily used, and crude tar is very indefinite. Undoubtedly a proper impregnation with A.R.E.A. No. 1 creosote is the most economical treatment in ultimate result. Aside from this grade of material, however, there is considerable difference of opinion as to the relative merits of using greater quantities of the Nos. 2 or 3 grades of straight creosote, or of using creosote-coal-tar combinations, or the proprietary solutions.

Since the electric companies are frequently also operators of gas plants, it is probable that there is a good commercial argument in favor of their developing the combination creosote-coal-tar treatment, even though theoretically the highest grade treatment is ultimately economical for absolutely permanent construction. In other words, the writer is of the opinion that, since such companies use large quantities of timber for installations the existence of which is not permanent, they would be justified in obtaining a less effective preservation at a lower cost. Further, if the development of their own resources would encourage their more general use of preservatives, the ultimate result would be advantageous, whether or not the kind of treatment for any particular case were absolutely the best possible.

Gas-house tar and coke-oven tar are practically alike chemically but differ greatly in the percentage of free carbon, the former having a comparatively high percentage and the latter usually a low percentage. High free-carbon tar, whether gas house or coke oven, should not be used as an addition to creosote oil. If, therefore, the combination material is to be used, only a low-carbon tar should be allowed in the creosote-coal-tar combination preservative.

Pressure Treatment.—Turning now from a discussion of the materials to be used in protecting poles from decay we find that several processes of treatment are in common use; the high-pressure, the open-tank, and the brush treatment, these being stated in the order of their effectiveness.

When treated by the high-pressure method, the timber is placed in metal cylinders and subjected to a steaming or heating process followed by a vacuum. After the vacuum has been maintained from one to two hours, the tank is completely filled with creosote oil and pressure is applied and maintained until the specified amount of creosote has been forced into the timber.

Open-tank Process.—The open-tank process consists in treating the butts of the poles only. Seasoned timber is immersed in a tank of hot preservative and kept there for a period of from one to three hours. The timber is then suddenly transferred to a bath at atmospheric temperature and kept there from one to three hours longer. By this process the air and moisture in the cells of the wood are first expanded and some driven off, while in the second bath the air and moisture contract drawing the preservative liquid into the timber.

Brush Treatment.—By the brush method, dry seasoned timber is given two or more coats of hot preservative applied with three- or four-knot rubber-set or wire-bound roofing brushes. The creosote should be heated to about 200°F. and kept at that temperature while being applied. A liberal quantity of liquid should be used and it should be well brushed into all crevices in the timber. Before applying preservative, the poles must be stripped of bark, inner skin, or dirt, and in fact should be scrubbed clean. Sufficient time should elapse between the application of the different coats for the preceding one to be absorbed; not less than one-day intervals are generally satisfactory. Poles should not be used for two or three days after treatment. In general it is more economical, both in labor and in the efficiency of the operation, to treat poles while they are in temporary storage, although some benefit is unquestionably derived from even a cold application at the site.

Whenever brush treatment is employed the entire butt should be coated up to about 2 ft. above the ground line. In addition, all crossarm gains, roofs and bolt holes should be painted with preservative.

SPECIFICATIONS FOR WOOD POLES

The purchaser shall have the right to make such inspection of the poles as may be desired. The inspector representing the purchaser shall have the power to reject any pole which is defective in any respect. Inspection, however, shall not relieve the contractor from the responsibility of furnishing proper poles.

Any imperfect poles which may be discovered before their final acceptance shall be replaced immediately upon the order of the purchaser, even though the defects may have been overlooked by the inspector.

Poles shall be subject to inspection by the purchaser, either in the woods where the trees are felled or at any point of shipment or delivery. Any poles failing to meet the requirements of these specifications may be rejected.

Seasoned poles shall have preference over green poles, provided they have not been held for seasoning long enough to have developed any of the timber defects hereinafter referred to. All poles shall be reasonably straight, well proportioned from butt to top, shall have both ends squared, the bark peeled and all knots and limbs closely trimmed.

DEAD POLES.—No dead poles and no poles having dead streaks covering more than one-quarter of their surface shall be accepted under these specifications. Poles having dead streaks covering less than one-quarter of their surface shall have a circumference greater than otherwise required. The increase in the circumference shall be sufficient to afford a cross-sectional area of sound wood equivalent to that of sound poles of the same class.

TWISTED, CHECKED OR CRACKED POLES.—No cracked poles, no poles containing large seasoning checks, and no poles having more than one complete twist for twenty (20) ft. in length shall be accepted under these specifications.

CROOKED POLES.—No poles having a short crook or bend, a crook or bend in two planes, or a reverse crook or bend, shall be accepted under these specifications. The amount of sweep measured between the six (6) ft. mark and the top of the pole shall not exceed one (1) in. for every six (6) ft. of length.

MISCELLANEOUS DEFECTS.—No poles containing sap rot, evidence of internal rot as disclosed by careful examination of black knots, hollow knots, woodpecker's holes, or plugged holes, and no poles showing evidences of having been eaten by ants, worms or grubs shall be accepted except that poles containing worm or grub marks below the six (6) ft. mark may be accepted.

CAT FACES.—No poles having "cat faces," unless the latter are small and perfectly sound and the poles have an increased diameter at the "cat face," and no poles having "cat faces" near the six (6) ft. mark or within ten (10) ft. of their tops shall be accepted.

WIND SHAKES.—No poles shall have cup shakes (checks in the form of rings) containing heart or star shakes which enclose more than ten (10) per cent. of the area of the butt.

BUTT ROT.—No poles shall have butt rot covering more than ten (10) per cent. of the total area of the butt. If butt rot is present it must be located close to the center in order that the pole may be accepted.

KNOTS.—Large knots, if sound and trimmed close, shall not be considered a defect. No poles shall contain loose, hollow or rotten knots.

DEFECTIVE TOPS.—Poles having tops of the required dimensions shall not be accepted unless the tops are sound. Poles having tops one (1) in. or more in excess of the required circumference may contain one (1) pipe rot not more than one-half (0.5) in. in diameter. Poles with double tops or double hearts shall be free from rot where the two parts or hearts join.

DEFECTIVE BUTTS.—No poles containing ring rot (rot in the form of a complete or partial ring) shall be accepted under these specifications.

Poles having hollow hearts may be accepted under the conditions shown in the following table:

TABLE 15

Average diameter of rot	Add to butt requirement (circumference)		
	25- and 30-ft. poles	35-, 40- and 45-ft. poles	50-, 55-, 60- and 65-ft. poles
2 in.	Nothing	Nothing	Nothing
3 in.	1 in.	Nothing	Nothing
4 in.	2 in.	Nothing	Nothing
5 in.	3 in.	1 in.	Nothing
6 in.	4 in.	2 in.	1 in.
7 in.	Reject	4 in.	2 in.
8 in.	Reject	6 in.	3 in.
9 in.	Reject	Reject	4 in.
10 in.	Reject	Reject	5 in.
11 in.	Reject	Reject	7 in.
12 in.	Reject	Reject	9 in.
13 in.	Reject	Reject	Reject

Scattered rot, unless it is near the outside of the pole, will be considered as being the same as heart rot of equal area.

DIMENSIONS.—The dimensions of the poles shall be not less than the values given in the following tables, the "top" measurement being the circumference at the top of the pole and the "butt" measurement the circumference six (6) ft. from the butt.

The dimensions specified for the six (6) ft. mark shall be required in all cases, but the top circumferences may differ from those shown in the following tables by not more than one-half (0.5) in. No pole shall be more than six (6) in. longer or three (3) in. shorter than the length for which it is accepted. If any pole is more than six (6) in. longer than is required it shall be cut back.

TABLE 16.—CHESTNUT[1]

Length of poles (ft.)	Circumferences of poles in inches					
	Classes					
	A		B		C	
	Top (in.)	6 ft. from butt (in.)	Top (in.)	6 ft. from butt (in.)	Top (in.)	6 ft. from butt (in.)
25	. . . .	. . . .	. . . .	. . . .	20	30
30	24	40	22	36	20	33
35	24	43	22	40	20	36
40	24	45	22	43	20	40
45	24	48	22	47	20	43
50	24	51	22	50	20	46
55	22	54	22	53	20	49
60	22	57	22	56	. . . .	. . . .
65	22	60	22	59	. . . .	. . . .
70	22	63	22	62	. . . .	. . . .
75	22	66	22	65	. . . .	. . . .
80	22	70	22	69	. . . .	. . . .
85	22	73	22	72	. . . .	. . . .
90	22	76	22	75	. . . .	. . . .

[1] Pole Size Tables 16, 17, 18, 19, 20, N.E.L.A. Handbook, 1914.

TABLE 17.—EASTERN WHITE CEDAR

Length of poles (ft.)	Circumferences of poles in inches					
	Classes					
	A		B		C	
	Top (in.)	6 ft. from butt (in.)	Top (in.)	6 ft. from butt (in.)	Top (in.)	6 ft. from butt (in.)
25	....	....	22	32	18¾	30
30	24	40	22	36	18¾	33
35	24	43	22	38	18¾	36
40	24	47	22	43	18¾	40
45	24	50	22	47	18¾	43
50	24	53	22	50	18¾	46
55	24	56	22	53	18¾	49
60	24	59	22	56		

TABLE 18.—WESTERN WHITE CEDAR, RED CEDAR, WESTERN CEDAR, IDAHO CEDAR

Length of poles (ft.)	Circumferences of poles in inches					
	Classes					
	A		B		C	
	Top (in.)	6 ft. from butt (in.)	Top (in.)	6 ft. from butt (in.)	Top (in.)	6 ft. from butt (in.)
20	28	30	25	28	22	26
22	28	32	25	30	22	27
25	28	34	25	31	22	28
30	28	37	25	34	22	30
35	28	40	25	36	22	32
40	28	43	25	38	22	34
45	28	45	25	40	22	36
50	28	47	25	42	22	38
55	28	49	25	44	22	40
60	28	52	25	46	22	41
65	28	54	25	48	22	43

SAWED REDWOOD POLES

The material desired under these specifications consists of poles of redwood (Sequoia Sempervirens) sawed to shape as hereinafter set forth.

QUALITY OF TIMBER AND WORKMANSHIP.—All poles shall be of sound No. 1 Common Redwood; and shall be reasonably straight and well sawn.

TABLE 19.—SAWED REDWOOD

Length of poles (ft.)	Classes			
	A		B	
	Top (in.)	Butt (in.)	Top (in.)	Butt (in.)
24	6 × 6	6 × 6	4 × 6	4 × 6
25	7 × 7	10 × 10	6 × 6	9 × 9
30	7 × 7	11 × 11	6 × 6	10 × 10
35	7 × 7	12 × 12	6 × 6	11 × 11
40	7 × 7	13 × 13	6 × 6	12 × 12
45	7 × 7	14 × 14	6 × 6	13 × 13
50	7 × 7	15½ × 15½	6 × 6	14 × 14

(Dimensions in inches)

The sectional dimensions of the sawn poles shall not be more than one-quarter (¼) in. under, or three-quarters (¾) in. over, the dimensions specified in the above table. No poles shall be more than three (3) in. longer or shorter than the lengths required in the above table.

SAPWOOD.—No poles shall have sapwood covering more than four (4) per cent. of the area of all the surfaces. No pole shall have sapwood for a distance of more than eight (8) ft. from the top. No sapwood shall be deeper than one (1) in. at any point.

KNOTS.—In 4″ × 6″ poles sound knots with a diameter smaller than one (1) in. may be present in any number. No 4″ × 6″ pole shall be accepted which contains in each five (5) superficial ft. more than one sound knot having a diameter of one (1) in. or more, or which contains any knots with a diameter greater than one and one-half (1½) in.

All other sizes of poles covered by these specifications may contain any number of sound knots with a diameter smaller than one and one-half (1½) in. No pole shall be accepted which contains in each five (5) superficial ft. more than one sound knot having a diameter of one and one-half (1½) in. or more, or which contains any knots of a diameter greater than two and one-half (2½) in.

NOTE.—Where diameters are specified in connection with knots a knot shall be rated on the basis of its average diameter.

SPECIFICATIONS FOR CREOSOTED YELLOW-PINE POLES

These specifications shall apply to Classes A, B and C poles of southern yellow pine treated with dead oil of coal tar.

QUALITY OF POLES.—All poles shall be sound southern yellow pine (longleaf, shortleaf, or loblolly yellow pine) squared at the butt, reasonably straight, well proportioned from butt to top, peeled and with knots trimmed close.   All poles shall be free from large or decayed knots. All poles shall be cut from live timber.

It is desired that all poles be well air-seasoned before treatment and such poles shall be treated in accordance with the requirements for treating seasoned timber contained in the "Specifications for Creosoting Timber" hereinafter referred to.   The poles shall not be held for seasoning, however, up to the point where local experience shows that sapwood decay would begin.   Unseasoned poles shall be treated in accordance with the requirements for treating unseasoned timber contained in the above-mentioned specifications.

All poles shall be sufficiently free from adhering inner bark before treating to permit the penetration of the oil.   If the inner bark is not satisfactorily removed when the pole is peeled, the pole shall be either shaved or allowed to season until the inner bark cracks and tends to peel from the pole.

DIMENSIONS.—The dimensions of the poles shall be not less than those given in the following table.

TABLE 20.—CREOSOTED YELLOW PINE

Length of pole (ft.)	Circumference of pole in inches					
	Classes					
	A		B		C	
	Top (in.)	6 ft. from butt (in.)	Top (in.)	6 ft. from butt (in.)	Top (in.)	6 ft. from butt (in.)
25	22	33	20	30	18	28½
30	22	35	20	32	18	30½
35	22	38	20	34	18	32
40	22	40	20	36	18	34
45	22	42½	20	38	18	36
50	22	44½	20	40	18	38
55	22	47	20	42½	18	40
60	22	49	20	44½	18	42
65	22	51	20	47	18	44
70	22	53	20	49	18	46
75	22	55	20	51	....	....
80	22	57	....	....	....	....

Framing of Poles.—Before the poles are treated with creosote they shall be framed, unless otherwise ordered, in the following manner and as shown in drawing No. (   ).

The top of each pole shall be roofed at an angle of ninety (90) degrees.

All Class A poles shall have eight (8) gains, all Class B poles shall have four (4) gains and all Class C poles shall have two (2) gains.

The gains shall be located on the side of the pole with the greatest curvature, and on the convex side of the curve. The faces of all gains shall be parallel.

Each gain shall be four and one-quarter (4¼) in. wide and one-half (½) in. deep and twenty-four (24) in. center to center. The center of the top gain shall be ten (10) in. from the apex of the gable. A twenty-one thirty-second ($2\frac{1}{32}$) in. hole shall be bored through the pole at the center of each gain perpendicular to the plane of the gain.

INSPECTION.—The quantity of dead oil of coal tar forced into the poles shall be determined by tank measurements and by observing the depth of penetration of the oil into the pole. If the poles have more than one and one-half (1½) in. of sapwood, the depth of penetration shall be not less than one and one-half (1½) in. If the sapwood is less than one and one-half (1½) in. thick, the dead oil of coal tar shall penetrate through the sapwood into the heartwood.

The depth of penetration shall be determined by boring the pole with a one (1) in. auger. The right is reserved to bore two holes at random about the circumference for this purpose, one hole to be five (5) ft. from the butt and one hole ten (10) ft. from the top. After inspection each test hole shall be filled first with hot dead oil of coal tar and then with a close-fitting creosoted wooden plug.

The rejection of any pole because of insufficient penetration shall not preclude its being retreated and again offered for inspection.

Design of Wood Poles.—If in Fig. 40, which represents a standard Class A chestnut pole, we assume for the present that all the bending loads are represented by one load, P of 1200 lb., 35 ft. from the ground, the bending moments and unit stresses at sections X2, X1, and at the ground line, will be as follows:

At X2,

$$M = 18.25 \text{ ft.} \times 12 \text{ in.} \times 1200 \text{ lb.} = 262{,}800 \text{ in.-lb.}$$
$$S = 0.0982 \times 12^3 = 169.69$$
$$F = M/S = 1550 \text{ lb. per square inch}$$

At X1,

$$M = 24.5 \times 12 \times 1200 = 352{,}800 \text{ in.-lb.}$$
$$S = 0.0982 \times 13.12^3 = 220.75$$
$$F = M/S = 1600 \text{ lb. per square inch}$$

At ground,

$$M = 35 \times 12 \times 1200 = 504{,}000 \text{ in.-lb.}$$
$$S = 0.0982 \times 15^3 = 331.42$$
$$F = M/S = 1530 \text{ lb. per square inch}$$

It should be noted therefore, that *the point of greatest stress is not necessarily at the ground line* but may be at some section above

the ground. If the pole under consideration were disproportion-
ally heavy at the butt, any computations made at the ground line
might be quite erroneous, although the difference in the example
given is negligible. This condition results from the fact that the
unit stress at any point depends on the distance from the load
and on the diameter of the pole at that point. Provided there
are no serious defects in a pole which may make some particular
point unusually weak it will, in theory, break at the point where
the diameter is 1.5 times the diameter at the point where the
load is applied. Therefore, poles may or may not fail at the

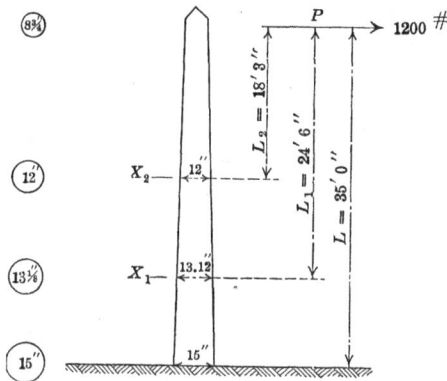

Fig. 40.

ground line depending on the taper. Further, if the butt
diameter exceeds the above critical diameter the pole may
experience some decay at the butt without becoming any weaker.

Referring to Fig. 41, if the following wires are to be carried,
with a minimum clearance of 30 ft., and a maximum stress in the
wires of 0.9 of the elastic limit, we have, if a 200-ft. span is
assumed,

One $\frac{3}{8}$-in. Siemens-Martin galvanized stranded steel.
Three No. 1 hard-drawn stranded copper, 33,000 volts.
Span 200 ft.
Normal sag = 1 ft. 3 in.   Normal tension   = 1020 lb.
Maximum sag = 2 ft. 0 in.   Maximum tension = 1960 lb.
Elastic limit, No. 1 cable = 2180 lb.
Wind pressure on wires:
$\frac{3}{8}$-in. ground wire = 0.917 lb. × 200 ft. = 183 lb.
No. 1 power wire = 0.885 lb. × 200 ft. = 177 lb.

The bending moment at the ground, for transverse loading, straight-line poles and no broken wires, is,

Ground wire, 183 lb. × 37 ft.   =   6,770 ft.-lb.
Power wires, 177 lb. × 34.5 ft.   =   6,100 ft.-lb.
Power wires, 177 lb. × 2 × 32 ft.   = 11,330 ft.-lb.
Wind on pole, 0.9 sq. ft. × 13 lb. × $\dfrac{37.5^2}{2}$ ft. =   8,225 ft.-lb.

Bending moment = 32,425 ft.-lb.

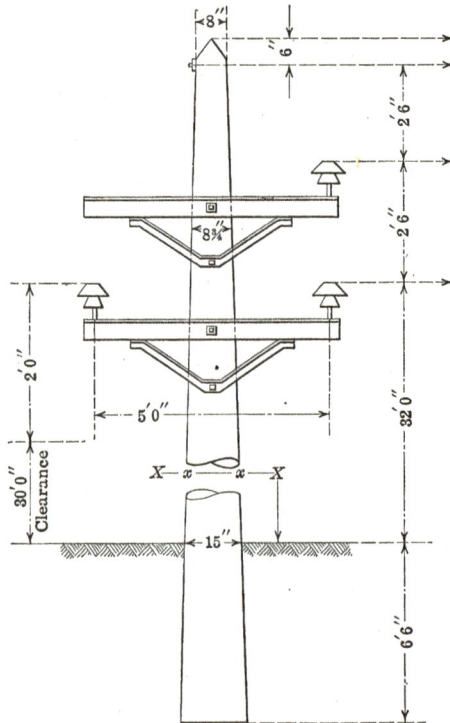

Fig. 41.—44-ft. pole.

The shear on the pole is,

Ground wire,   183 lb.
Three power wires,   531 lb.
Wind on pole,   438 lb.

Total shear,   1152 lb.

and

$$\frac{32,425}{1152} = 28.1$$

Or the load is equivalent to a single load, $P = 1152$ lb., 28.1 ft. above the ground.

If the weakest section of the pole is assumed as being at the ground line, which is usually *not* correct, the unit stress under the first condition of loading is,

$$M = \frac{SI}{c}$$

$$S = \frac{32,425 \text{ ft.-lb.} \times 12 \text{ in.}}{\dfrac{I}{c}}$$

$$\frac{I}{c} = \frac{1}{10} \times \text{diam.}^3 \text{ (approx.)} = \frac{15^3}{10} = 337.5$$

$S = 1150$ lb. per square inch, bending stress.

To obtain the maximum unit bending stress in the pole:
Center of gravity of wire loads below top of pole:

$$183 \text{ lb.} \times 6 \text{ in.} = 1{,}098 \text{ in.-lb.}$$
$$177 \text{ lb.} \times 36 \text{ in.} = 6{,}372 \text{ in.-lb.}$$
$$354 \text{ lb.} \times 66 \text{ in.} = 23{,}364 \text{ in.-lb.}$$

714 lb.   30,834 in.-lb.
30,834 in.-lb. ÷ 714 lb. = 43 in. below top of pole.

The location of the point of maximum stress below the center of gravity of wire loads can be found from the formula, page 75:

$$y = \frac{2P}{w} - \frac{d}{t} + 2a \pm \sqrt{\frac{4P^2}{w^2} - \frac{2Pd}{wt} + \frac{8Pa}{w} + \frac{d^2}{t^2} - \frac{2da}{t} + a^2}$$

In this case

P = wind on wires = 714 lb.
w = wind per inch of pole = 1 lb.
d_1 = diam. of pole at top = 7.4 in.
d_2 = diam. of pole 6 ft. 0 in. above butt = 15 in.
d_3 = diam. of pole at point of maximum stress.
t = increase in diameter per inch of length
$$= \frac{15 - 7.4}{38 \text{ ft.} \times 12} = \frac{1}{60}$$
d = diam. of pole at load P = 8 in.
a = dist. of load P below top of pole = 43 in.

Substituting these in the above formula:

$$y = 304 \text{ in.} = 25 \text{ ft. } 4 \text{ in.}$$

The maximum stress occurs, therefore, 25 ft. 4 in. below the center of gravity of the wire loads, or 28 ft. 11 in. below the top of the pole.

$$d_3 = d + ty = 8 \text{ in.} + \frac{1}{60} \times 304 = 13 \text{ in.}$$

Maximum stress in pole

$$S = \frac{M}{\dfrac{I}{c}} = \frac{Py + \frac{\omega}{2}(y + a)^2}{0.098d_3{}^3}$$

$$= \frac{714 \times 304 + \dfrac{1}{2}(304 + 43)^2}{0.098 \times 13^3}$$

$$= \frac{277,260}{215.3} = 1290 \text{ lb. per square inch.}$$

As the breaking strength for a chestnut pole is 5100 lb. per square inch, the factor of safety is $\dfrac{5100}{1290} = 4.$

The stress at the ground line equals

$$s = \frac{714 \times (450 - 43) + \dfrac{1 \times 450^2}{2}}{0.098 \times (14.9)^3}$$

$$= \frac{391,848}{324.2} = 1200 \text{ lb. per square inch.}$$

The bending moment at the ground for a pole at a 5° corner, is found as follows:

Maximum wire tension = 1960 lb.

Component due to corner (Fig. 32) = 0.10 tension

or

$$1960 \text{ lb.} \times 0.10 = 195 \text{ lb. per wire.}$$

Wind on wires and pole

(same as before)	= 32,425 ft.-lb.
195 lb. × 37 ft.	= 7,215 ft.-lb.
195 lb. × 34.5 ft.	= 6,725 ft.-lb.
195 lb. × 2 × 32 ft.	= 12,480 ft.-lb.
Total bending moment	= 58,845 ft.-lb.

1152 lb. wind on wires and pole
780 = 195 × 4 wires, corner loading
——————
1932 Shear

and

$$\frac{58,845}{1932} = 30.4 \text{ ft.}$$

Fig. 42.—Corners on single pin insulators.

Therefore

$P = 1932$ lb., equiv. load 30.4 ft. above ground

$$S = \frac{58,845 \times 12}{337.5} = 2100 \text{ lb. per sq. in.}$$

In Fig. 42 is shown a one-circuit wood-pole line, in which the wires turn rather sharp corners on single-pin insulators, a practice which is objectionable. The illustration shows considerable right-of-way clearing, but it appears that the poles will presumably not be subjected to very severe wind loads on account of the shelter afforded by adjoining timber. It is also evident that a very wide clearing would be necessary to entirely protect the line from adjoining trees.

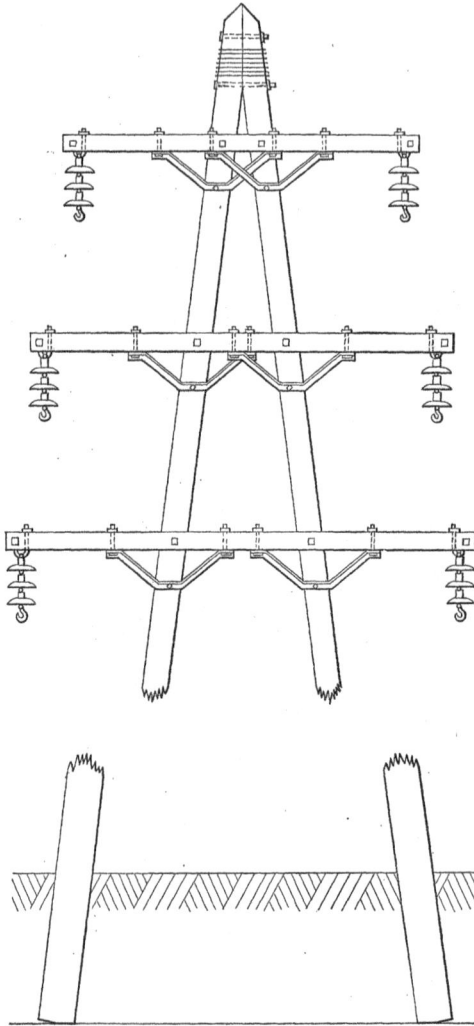

Fig. 43.—Design for wooden A-frame.

A-frames and H-frames.—Timber A-frames composed of two poles spliced together at the top and with their butts separated transverse to the line are useful chiefly where large timber is expensive, as such construction permits the use of slender poles, one of which would not have sufficient strength. These frames have not been used to any considerable extent, however, in this country. In the direction of the line, the strength is twice that of the single poles, while it is considerably greater in a transverse direction, the amount depending largely on the bracing provided

FIG. 44.—One-circuit H-frame.

to prevent buckling of each pole. Except at corners, these frames are relatively too strong across the line as compared with their strength in the direction of the line.

The H-frame, on the other hand, while having less theoretical strength across the line, is a useful type of construction, particularly for heavy lines in bad ground. Its width at the top permits a larger number of wires per crossarm, while utilizing the strength of the arms as simple beams instead of cantilevers.

In Fig. 44 is shown a one-circuit H-frame, consisting of two light timber poles. This is typical of the characteristic usefulness

7

of an H-frame in that two slender poles can be used to provide
adequate strength.   On account of their strength H-frames

Fig. 45.—H-frame crossing, metal grounding arms.

may also be employed to support heavy lines, although they
are more frequently used for heavy telephone and telegraph
trunk lines than for transmission lines (Fig. 86).

TABLE 20a.—WOODEN POLE LINES

Lines	Voltage	Miles	Standard span (ft.)	Conductors			Ground wire		Telephone		Protective treatment	Estimated ave. life (yrs.)	Design loading
				No.	Size	Material	Size	Material	Size	Material			
O. P. & W. Co.	66,000	120	150	3	0 / 00	Sol. Cop. / Str. Cop.	6	BB. Galv. Iron	10	Cop.	Brush		
P. L. & P. Co.	66,000	280 / 60	300 / 300	3	1-00 / 2-0000 / 1	Str. Cop. / Str. Al. / Str. Cop.	......	None	8 / 10	Cop. clad Iron	Brush	9	
K. G. & E. Co.	60,000	......	125	...	......	Cop.	¼	Galv. Steel	14	Iron	Butt treat.	15	
V. I. P. Co.	60,000	74	300	3	00	Str. Al.	......	None	10	Galv. Iron	Butt treat.	15	½"-8.0 lb.
T. P. & L. Co.	60,000	37	300	3	1	Str. Cop.	⅜	Galv. str. Steel	......	None	Pressure treat.	......	½"-10.5 lb
S. J. L. & P.	60,000	704	350	3	0 / 000 / 0000	Cop. and Al.	......	None	8	Cop.	Butt treat.		
C. P. & L. Co.	60,000	128	300	3	2 / 1 / 0	Str. Cop.	⅜	S.M. Steel	8	Iron	Brush	7	½"-50 M.P.H.
U. P. & L. Co.	45,000	285	180 / 240 / 300	3	0 / 2-4-5 / 3 / 6-9	Str. Cop. / Sol. Cop. / Al. Iron	......	None	......	None	Brush	15	
T. T. M. Co.	44,000	66	200	3	4	Cop.	......	None	10	Cop.	Butt treat.		
S. P. & E. Co.	44,000	290	132	3	2 / 00 / 0000	Str. Al.	¼	Str. Steel	9 B. W.G.	B.B.Galv. Iron			
C. P. & L. Co.	44,000	51	140	3	4 / 6	Cop.	......	None	10	E.B.B. Galv. Iron	......	......	½"-60 M.P.H

TABLE 20a.—WOODEN POLE LINES. *Continued*

Lines	Voltage	Miles	Standard span (ft.)	Conductors			Ground wire		Telephone		Protective treatment	Estimated ave. life (yrs.)	Design loading
				No.	Size	Material	Size	Material	Size	Material			
D. P. & T. Co.	40,000	145	90–150	3	000	Str. Cop. Hemp Core	6	....	....	....	....	12	½"–100 M.P.H.
C. C. P. & L. Co.	38,000	90	100	3 6	2	Cop.	....	None	12	Cop.	Brush	10–20	½"
E. D. E. Co.	33,000	125	150	3	4 2 0	Sol. Cop.	6	BB. Galv. Iron	10	Cop.	2-Brush	18–Cedar	
K. U. Co.	33,000	70	150	3	1 6	Cop.	14	Galv. Steel	8	Cop. clad	Butt treat.		
A. E. P. Co.	30,000	55	100	6	030	Sol. Cep	1	Solid Galv. steel	12	Cop.	(treating) now		
Y. R. P. Co.	22,000	40	225	3	4 1	Sol Cop. Str. Cop	$\frac{3}{8}$	S.M. Steel	8	Iron	Open tank	....	½"–50 M.P.H.
C. I. L. & P. Co.	22,000	72 103	200 100	3	2 4	Cop. (20 mi. Al. equiv.)	$\frac{5}{16}, \frac{7}{16}$	S.M. str. Galv. Steel	10	Cop. clad	2-Brush after erection	4–Cypress 8–Chestnut 20–Cedar (treated)	
M. S. R. & L. Co.	22,000	102	120	6	0 4	Str. Cop. Sol. Cop.	4	Cop.	12	Cop. and Cop. clad	None	14–Chestnut 20–Cedar	¼"–8.0 lb.
W. L. Co.	13,200	46	120	....	2-00	Cop., bare and t.b.w.	....	None	.....	None	Butt treat. Dipped arms	8–15	
E. St. L. & S.	13,200	97	100	6	2 2	Cop. (15 mi. Al. equiv.)	2	Cop. clad	12	B.B.Iron t.b.w.	Brush	6–Cypress 15–Chestnut 15–Cedar	

TABLE 20a.—WOODEN POLE LINES. *Continued*

Lines	Voltage	Miles	Standard span (ft.)	Conductors			Ground wire		Telephone		Protective treatment	Estimated ave. life (yrs.)	Design loading
				No.	Size	Material	Size	Material	Size	Material			
P. S. E. Co.	13,000	75	100 125	3-15	4- 250,000	Cop. (str. above 00)	3/8	Cop. clad (on a few lines)			Will adopt brush treat.	13-Chestnut	
E. E. Co. & C. T. Co.	11,500	200	75 125	....	3	Cop.	1/4	S.M. Galv. Steel	....	None	Experimenting	15	1/2"-No wind 60 M.P.H.-No ice
U. R. & P. Co.	11,000	36	125	3 6	4-00	Sol. Cop.	5/16	Galv. Steel	....	None	Pressure	20	
I. O. L. & P. Co.	66,000	48	225	3	1	Str. Cop.	6	B.B. Iron	8	Cop.	Brush	8	No ice-8.0 lb.
	66,000	54	225	3	1	Str. Cop.	1/4	S.M. str. Galv. Steel	10	Cop. clad	Brush	....	No ice-8.0 lb.
	66,000	52	300	3	131,100 65,300	Str. Al.	6	B.B. Iron	10	Cop.	Brush	....	No ice-8.0 lb.
	44,000	27	225	3	1	Str. Cop.	....	None	10	Cop.	Brush	....	No ice-8.0 lb.
	23,000	31	175	3	4	Str. Cop.	....	None	8	Galv. Iron	Brush	....	No ice-8.0 lb.
	23,000	50	140	3	6	Str. Al. Cop.	....	None	8	None	Brush	....	No ice-8.0 lb.
	11,500	26	200	3	4	Cop.	....	None	....	None	Brush	....	
M. R. P. Co.	66,000	27	300	3	2	Str. Cop.	2-5/16	S.M. Galv. Steel	6	Cop clad	2-Coal tar	....	1/2"-8.0 lb.
	33,000	32	140	3	2	Str. Cop	1/4	S.M. str. Galv. Steel	8 10	Galv. Iron Cop. clad	Water gas drippings	....	1/2"-8.0 lb.

TABLE 20a.—WOODEN POLE LINES. *Continued*

Lines	Voltage	Miles	Standard span (ft.)	Conductors			Ground wire		Telephone		Protective treatment	Estimated ave. life (yrs.)	Design loading
				No.	Size	Material	Size	Material	Size	Material			
M. R. P. Co.	11,000	36	140	6	......	Str. Al. (0000 equiv.)	5⁄16	S.M. str. Galv. Steel	8	Cop.	Butt Creosote	15	½"—8.0 lb.
W. C. P. Co.	50,000 17,000 10,000	270	100-240	3	1-6 300,000	Str. Cop. Al. (hemp core)	2	Galv. Iron	8 10	Galv. Iron	Butt treat. (within 4 yrs.)	15-20	
P. L. & P. Co.	60,000 50,000 15,000 10,000	84 262 301 33	......	3-6	6- 250,000	Cop., Al. Steel Core Al.	¼	Str. Steel Galv. on 31 mi.—50,000	14-9	Steel, Cop. clad, Cop.	Generally tank treated since 1914	10-15	—8.0 lb.
N. E. L. Co.	33,000 11,000	33 22	100 100-120	6	4 0	Cop. Cop. t.b.w.	......	None	.....	None	Brush		
W. E. Co.	66,000 33,000	48	120-150	3	1	Str. Cop.	3⁄8 4	S.M. Steel Cop. clad	8	Cop. clad	Butt treat.	......	½"—8.0 lb.
S. C. E. Co.	30,000 15,000 10,000	1,000	150-300	3 6	6- 0000	Cop.	.....	None	10	Cop.	Open tank	20	
T. & N. P. Co.	25,000 12,500	40	100 125 150	3 6	.....	Cop.	5⁄16	Str. Steel	12	Cop.	2-Brush		
I. P. Co.	13,200	70 22	110 140	3 3	4 6	Sol. Cop.	9 B.W.G.	Galv. Steel	10	Galv. Steel	....	13	
	33,000	110	200	2	8	Cop. clad	......	.....	.....	.....	Open tank	......	1"—60 M.P.H. lb.

CHAPTER VI

STEEL POLES AND TOWERS

OPEN-HEARTH STEEL

Manufacturers' Standard Specifications. Issue of Feb. 6, 1914.
(Abstracts)

All tests and inspections shall be made at the place of manufacture prior to shipment.

The tensile strength, limit of elasticity and ductility shall be determined from a standard test piece cut from the finished material.

The elongation shall be measured on an original length of 8 in. Rivet rounds and small bars shall be tested full size as rolled.

Two tests pieces shall be taken from each melt or blow of finished material—one for tension and one for bending—but in case either test develops flaws, or the tensile test piece breaks outside of the middle third of its gaged length, it may be discarded and another test piece substituted therefor.

Material which is to be used without annealing or further treatment shall be tested in the condition in which it comes from the rolls. When material is to be annealed or otherwise treated before use, the specimen representing such material shall be similarly treated before testing.

Finished bars shall be free from injurious seams, flaws or cracks, and have a workmanlike finish.

MAXIMUM PHOSPHORUS.—0.10 per cent.

RIVET STEEL.—Ultimate strength, 48,000 to 58,000 lb. per square inch. Elastic limit, not less than one-half the ultimate strength. Percentage of elongation, $\dfrac{1,400,000}{\text{ultimate strength}}$. Bending test, 180° flat on itself, without fracture on outside of bent portion.

RAILWAY BRIDGE GRADE.—Ultimate strength, 55,000 to 65,000 lb. per square inch. Elastic limit, not less than one-half the ultimate strength. Percentage of elongation, $\dfrac{1,400,000}{\text{ultimate strength}}$. Bending test, 180° to a diameter equal to thickness of piece tested, without fracture on outside of bent portion.

MEDIUM STEEL.—Ultimate strength, 60,000 to 70,000 lb. per square inch. Elastic limit, not less than one-half the ultimate strength,

Percentage of elongation, $\dfrac{1,400,000}{\text{ultimate strength}}$. Bending test, 180° to a diameter equal to thickness of piece tested, without fracture on outside of bent portion.

For material less than $\frac{5}{16}$ in. and more than $\frac{3}{4}$ in. in thickness, the following modifications shall be made in the requirements for elongation:

For each decrease of $\frac{1}{16}$ in. in thickness below $\frac{5}{16}$ in., a deduction of $2\frac{1}{2}$ per cent. shall be made from the specified elongation.

In rounds of $\frac{5}{8}$ in. or less in diameter, the elongation shall be measured in a length equal to eight times the diameter of section tested.

The variation in cross-section or weight of more than $2\frac{1}{2}$ per cent. from that specified will be sufficient cause for rejection, except in the case of sheared plates, which will be covered by certain permissible variations.

Rivets and Bolts.—A well-made hot-driven rivet is superior to even a turned bolt because the excess length of the rivet has been upset into contact with the sides and irregularities of the hole, and because in cooling the contraction of the rivet presses the riveted pieces together and develops a friction which increases the strength of the joint. The use of turned or machined bolts in drilled or accurately reamed holes would be of prohibitive cost for pole or tower work, while drawn-wire bolts (which are truly circular), in holes enlarged with a conical reamer, are not worth the additional expense. Since the number of bolts per joint is small, rarely over two and frequently but one, it is probable that the friction between the connected pieces is negligible.

For transmission line construction the strength of a rivet or bolt is either its shearing or its bearing value. The former is the shearing strength per square inch of the rivet or bolt material multiplied by the area of the cross-section, and assuming, as will usually be the case, that 48,000–58,000-lb. material is used, we have the following shearing values:

TABLE 21.—SHEARING VALUES OF RIVETS AND BOLTS

Diameter rivet or bolt (in inches)	Ultimate shear (in pounds) (36,000 lb. per square inch)	Working values (in lb.)	
		Shop rivets (15,000 lb. per square inch)	Field rivets or bolts (12,000 lb. per square inch)
$\frac{1}{2}$	7,000	2,900	2,400
$\frac{5}{8}$	11,000	4,600	3,700
$\frac{3}{4}$	16,000	6,600	5,300
$\frac{7}{8}$	21,500	9,000	7,200

The bearing value of a rivet or bolt is that of the effective area of metal pressed together during the transfer of stress, and, therefore, equals the product of the diameter of the rivet or bolt, the thickness of the thinner riveted piece, and the unit bearing value of the material.

TABLE 22.—ULTIMATE BEARING

Diameter rivet or bolt (in inches)	Ultimate bearing (72,000 lb. per square inch)					
	Thickness of thinner connected piece (in inches)					
	$\frac{1}{8}$	$\frac{3}{16}$	$\frac{1}{4}$	$\frac{5}{16}$	$\frac{3}{8}$	$\frac{7}{16}$
$\frac{1}{2}$	4,500	6,700	9,000	11,300	13,500	15,700
$\frac{5}{8}$	5,600	8,400	11,200	14,000	16,800	19,600
$\frac{3}{4}$	6,700	10,100	13,500	16,800	20,200	23,600
$\frac{7}{8}$	7,900	11,800	15,800	19,700	23,600	27,600

TABLE 23.—BEARING VALUES OF RIVETS AND BOLTS[1]

Diameter rivet or bolt (in inches)	Working values											
	Shop rivets (30,000 lb. per square inch)						Field rivets or bolts (24,000 lb. per square inch)					
	Thickness of thinner connected piece (in inches)											
	$\frac{1}{8}$	$\frac{3}{16}$	$\frac{1}{4}$	$\frac{5}{16}$	$\frac{3}{8}$	$\frac{7}{16}$	$\frac{1}{8}$	$\frac{3}{16}$	$\frac{1}{4}$	$\frac{5}{16}$	$\frac{3}{8}$	$\frac{7}{16}$
$\frac{1}{2}$	1,800	2,800	3,700	4,700	5,600	6,500	1,500	2,200	3,000	3,700	4,500	5,300
$\frac{5}{8}$	2,300	3,500	4,700	5,900	7,000	8,200	1,800	2,800	3,700	4,700	5,600	6,500
$\frac{3}{4}$	2,800	4,200	5,600	7,000	8,400	9,800	2,200	3,300	4,500	5,600	6,700	7,900
$\frac{7}{8}$	3,200	4,900	6,500	8,200	9,800	11,500	2,600	3,900	5,300	6,600	7,900	9,200

In bridge and building construction, it is customary to specify that the distance between rivet holes shall be not less than three times the diameter of the rivet, and the distance from the center of the hole to the end or edge of the piece shall be not less than one and one-half times the diameter of the rivet. In practice, however, particularly when using small thin sections, these minimum distances are often reduced. In transmission line structures, the stresses to be transferred from the bracing to the main members are usually small and there is little danger of the material failing between or outside the holes, provided an excess-

[1] In the working values for $\frac{1}{8}$-in. material, a more conservative factor of safety has been used, since the theoretical ultimate values are not always obtained in practice. For instance, it has been found by test that $\frac{1}{8}$-in. material may crumple, allowing the bolt to pull through the hole at a stress less than the theoretical ultimate bearing value.

ively close spacing is prohibited. The distance from a hole to a rolled edge may be made slightly smaller than that to a sheared edge or end, since the material of the former is free from any injury due to the shearing process.

In Fig. 46 are shown the minimum spacing, edge, and end distances, for each size of rivet or bolt, below which further reduction is inadvisable. Where clearance will permit, the end distances shown should be increased about $\frac{1}{8}$ in. to $\frac{1}{4}$ in. From

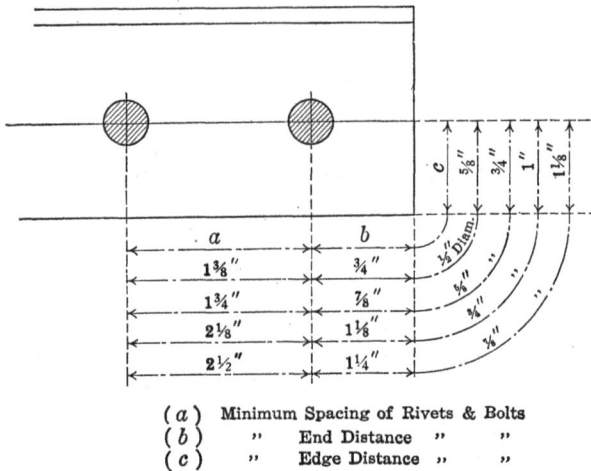

(a) Minimum Spacing of Rivets & Bolts
(b)   ,,   End Distance   ,,   ,,
(c)   ,,   Edge Distance ,,   ,,

Fig. 46.—Minimum spacing, edge, and end distances.

the edge distances given and assuming the usual thickness of material and sizes of nut, the minimum section of angle to be used for each diameter of bolt is found to be:

Table 24.—Minimum Angle Sections

Diameter rivet or bolt (in inches)	Minimum angle
$\frac{1}{2}$	$1\frac{1}{2}$ in. L
$\frac{5}{8}$	$1\frac{3}{4}$ in. L
$\frac{3}{4}$	$2\frac{1}{4}$ in. L
$\frac{7}{8}$	$2\frac{1}{2}$ in. L

Lacing.—The function of lacing is to stiffen the connected members by reducing the unsupported length of the compression section and also to transmit shearing stresses. If the shear is relatively large, the limiting condition may be the number of rivets connecting the lacing to the main section, otherwise it will be the stiffness of the lacing itself. That is, the lacing is a compression member whose strength depends on its ratio of stiffness, or l/r. Since the minimum radius of gyration of a flat or bar is much smaller than that of an angle, the unsupported length of the former must be less. Again, flat lacing is more subject to accidental injury than angle lacing because a slight bend in the direction of the thickness may easily occur and make the theoretical compressive strength negligible.

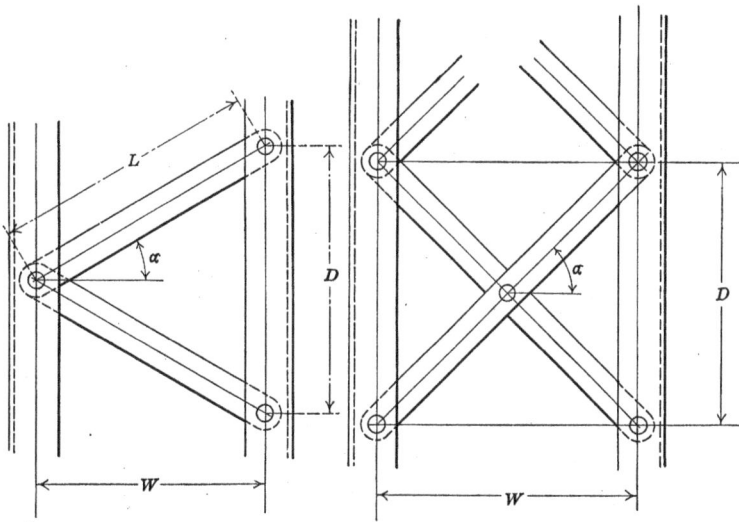

FIG. 47.—Single flat lacing.   FIG. 48.—Double flat lacing.

The value of r of a lacing bar is approximately 0.3 of its thickness; therefore, increased stiffness can be obtained only by additional thickness or by reducing the length.

When double lacing is used, some reduction in effective length may be assumed because of the connection at the intersection. With flat lacing, however, it is not correct to assume that the effective length is the distance from the end hole to the intersection.

The inclination or angle α of the lacing affects both the length of the bars and their stress. The compressive stress in lacing is:

$$C = \text{shear } S \times \text{sec. } \alpha$$

or

$$C = 1.155 \ S \text{ for } \alpha = 30°$$
$$C = 1.414 \ S \text{ for } \alpha = 45°$$
$$C = 2.000 \ S \text{ for } \alpha = 60°$$

Therefore, both the length of the bar and the stress will be increased 73 per cent. by increasing the angle α from 30° to 60°. The reduced strength and increased stress may cause either the thickness of the bar or the strength of the rivet to become the limiting condition.

In addition, it is unwise to use values of α much in excess of 30° for single bars, 45° for double bars, and 45° for angles, as the stiffening effect of a light member connected by one rivet is very small with excessive inclinations.

Angle Lacing.—Owing to the fact that the radius of gyration of an angle is larger than that of a flat, the former section allows a considerable increase in the width of the main members with less material in the lacing. The angle section may depend on the size of the bolt needed to transmit the stress in the lacing or, if the latter is turned in, on the permissible end and edge distances.

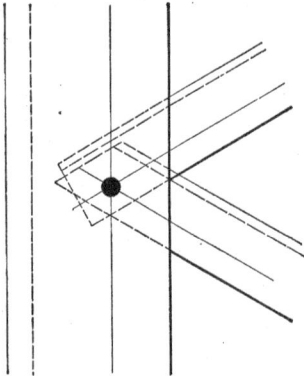

Fig. 49.—Single angle lacing.

In Table 25 will be found the values of r for various sections, and the length corresponding to different ratios of l/r. The maximum permissible value of l/r will depend to some extent on the character of the service expected of the bracing, as well as on its position in the structure. Bracing for secondary members, which are not liable to accidental injury or torsion, may be allowed larger ratios than main compression members, which from their position may be subject to both injury and torsion. Again, in selecting angle sections and the ratio l/r for any member, some consideration should be given to its position and protective coating, as the effect of these may make it advisable to employ a thicker angle. Thus, a bracing angle with the outstanding flange turned in and upward, and the angle itself in a vertical or inclined plane, is less subject to injury either from accident or corrosion than a similar angle reversed or in a horizontal plane.

TABLE 25.—ANGLES

Section	Area	Weight	Least r	60	80	100	120	150	180	220
				Length in inches corresponding to various values of *l/r*						
1½×1¼×3/16	0.48	1.6	0.26	....	....	....	....	....	....	....
×¼	0.63	2.1	0.26	....	....	....	....	....	....	....
1½×1½×1/8	0.36	1.2	0.30	18	27	30	36	45	54	66
×3/16	0.53	1.8	0.29	17	23	29	35	44	52	64
×¼	0.69	2.3	0.29	17	23	29	35	44	52	64
1¾×1¼×1/8	0.36	1.2	0.27	....	....	....	....	....	....	....
×3/16	0.53	1.8	0.27	....	....	....	....	....	....	....
×¼	0.69	2.3	0.27	....	....	....	....	....	....	....
1¾×1¾×1/8	0.42	1.4	0.35	21	28	35	42	52	63	77
×3/16	0.63	2.1	0.34	20	27	34	41	51	61	75
×¼	0.82	2.8	0.34	20	27	34	41	51	61	75
2 ×1¼×3/16	0.57	2.0	0.27	17	23	29	35	44	52	64
×¼	0.75	2.6	0.27	17	23	29	35	44	52	64
2 ×2 ×3/16	0.72	2.4	0.39	23	31	39	47	58	70	86
×¼	0.94	3.2	0.39	23	31	39	47	58	70	86
2¼×2¼×3/16	0.81	2.8	0.44	26	35	44	53	66	79	97
×¼	1.07	3.6	0.44	26	35	44	53	66	79	97
2½×1½×3/16	0.72	2.4	0.33	20	26	33	40	49	59	73
×¼	0.94	3.2	0.32	19	26	32	38	48	58	70
2½×2 ×3/16	0.81	2.8	0.43	26	34	43	52	64	77	95
×¼	1.07	3.6	0.42	25	34	42	50	63	76	92
2½×2½×3/16	0.91	3.1	0.49	29	39	49	59	73	88	108
×¼	1.19	4.1	0.49	29	39	49	59	73	88	108
×5/16	1.47	5.0	0.49	29	39	49	59	73	88	108
2¾×2¾×3/16	1.00	3.4	0.54	32	43	54	65	81	97	119
×¼	1.32	4.5	0.54	32	43	54	65	81	97	119
×5/16	1.63	5.6	0.54	32	43	54	65	81	97	119
3 ×2 ×3/16	0.91	3.1	0.44	26	35	44	53	66	79	97
×¼	1.19	4.1	0.43	26	34	43	52	64	77	95
×5/16	1.47	5.0	0.43	26	34	43	52	64	77	95
3 ×2½×3/16	1.00	3.4	0.53	32	42	53	64	79	95	117
×¼	1.32	4.5	0.53	32	42	53	64	79	95	117
×5/16	1.63	5.6	0.53	32	42	53	64	79	95	117
3 ×3 ×3/16	1.09	3.7	0.59	35	47	59	71	88	106	130
×¼	1.44	4.9	0.59	35	47	59	71	88	106	130
×5/16	1.78	6.1	0.59	35	47	59	71	88	106	130
3½×2½×¼	1.44	4.9	0.54	32	43	54	65	81	97	119
×5/16	1.78	6.1	0.54	32	43	54	65	81	97	119
3½×3 ×¼	1.56	5.4	0.63	38	50	63	76	94	113	139
×5/16	1.94	6.6	0.63	38	50	63	76	94	113	139
3½×3½×5/16	2.09	7.2	0.69	41	55	69	83	103	124	152
×3/8	2.49	8.5	0.68	41	54	68	82	102	122	150
×7/16	2.88	9.8	0.68	41	54	68	82	102	122	150
×½	3.25	11.1	0.68	41	54	68	82	102	122	150
4 ×4 ×5/16	2.41	8.2	0.79	47	63	79	95	118	142	174
×3/8	2.86	9.8	0.79	47	63	79	95	118	142	174
×7/16	3.31	11.3	0.78	47	62	78	94	117	140	172
×½	3.75	12.8	0.78	47	62	78	94	117	140	172
5 ×5 ×3/8	3.61	12.3	0.99	59	79	99	119	148	178	218
×7/16	4.19	14.3	0.98	59	78	98	118	147	176	216
×½	4.75	16.2	0.98	59	78	98	118	147	176	216
6 ×6 ×3/8	4.36	14.9	1.19	71	95	119	143	178	214	262
×7/16	5.06	17.2	1.19	71	95	119	143	178	214	262
×½	5.75	19.6	1.18	71	94	118	142	177	212	260
×9/16	6.44	21.9	1.18	71	94	118	142	177	212	260

Tower Connections.—As it is usually advantageous, from a manufacturing standpoint, to maintain like punching on both flanges of the main leg angles, the panel points are often at the same elevation on all four sides of the tower and the holes are

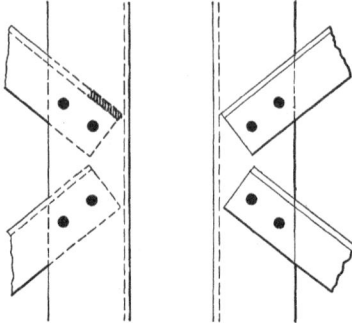

FIG. 50.—Bracing connection.

opposite each other. This necessitates clipping the outstanding flange of one brace at each panel point, to clear the inside brace on the adjoining face and also to provide space for the insertion of the inside connection bolt (Fig. 50).

An alternative method is to stagger the main panel points a few inches and thus obtain the necessary clearances without clipping. To do this, and also maintain like bracing angles on all faces of the tower, it is necessary to make the tower out of square, so that the increased width of two opposite faces will compensate for the greater length of the diagonal bracing in its lowered position.

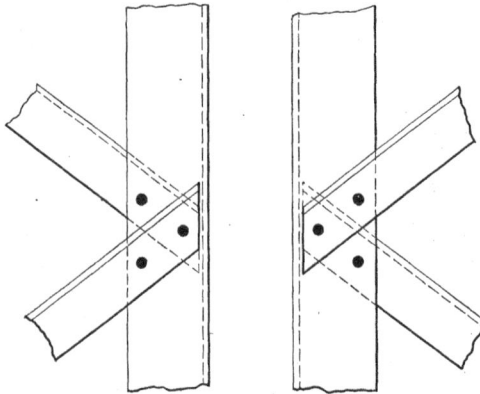

FIG. 51.—Bracing connection.

The outstanding flanges of horizontal or inclined angles should always be turned up, as in this position they drain and dry quickly and do not collect dirt or hold water. For similar reasons it is inadvisable to use any closed pockets, or semi-closed pockets, anywhere in the structure, as they are certain to become clogged with refuse and filled with water. Since moisture is a necessary condition of all decay and corrosion, rapid and thorough drainage are essential to a good design whether the material be timber or steel.

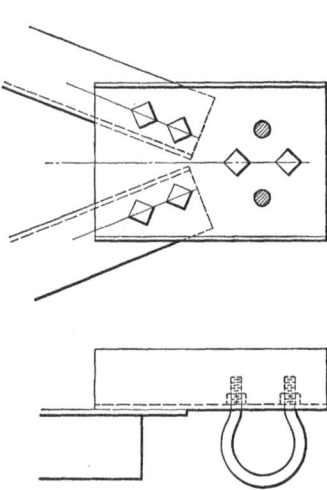

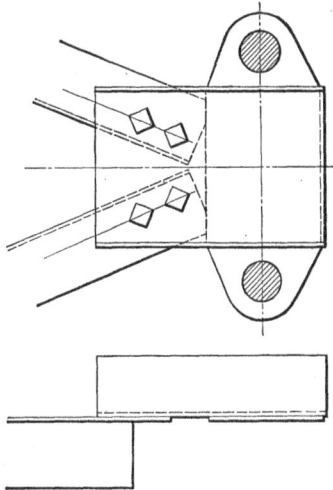

Fig. 52.—Suspension insulator connections.   Fig. 53.—Strain insulator connections.

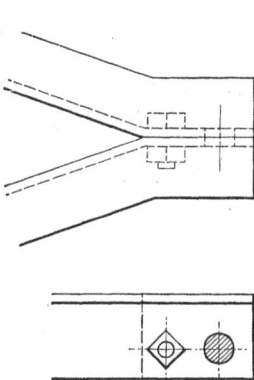

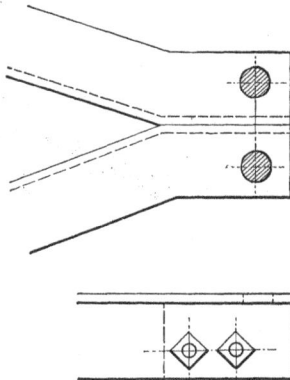

Fig. 54.   Fig. 55.

Single ½-in. bolt connections should be prohibited in the main bracing system of wide-base towers, except possibly for the connection of secondary members such as sub-panel struts, whose sole function is to reduce the unsupported length of other members. Such struts have frequently been given odd inclinations as compared with the main diagonal system, with a result far from pleasing in appearance. It is generally true that the diagonal bracing on tangent towers with well-inclined main legs has to withstand only relatively low stresses. In the frantic endeavor to reduce costs, this fact has led to the use of widely separated bracing, which is incapable of properly supporting the most important section—the main leg members. In other cases, in an effort to obtain the greatest possible theoretical strength, the bracing has been closely spaced, but cut down in section to such members as 1¼ in. × 1¼ in. × ⅛-in. angles with large values of $\dfrac{l}{r}$.

Fig. 56.

The connection of disc-type insulators to steel poles or towers is usually made by providing a U-bolt or plate into which a hook is inserted. In other cases, the connection hole is made in the crossarm angles as shown in Fig. 54.

There is no particular merit in one form of connection rather than in another, provided the thickness of the crossarm material is sufficient to transmit the stress to the arms. The thickness of the material outside the hook hole must be about ⅝ in. on account of the spread of the hook and the inside edges of the hole should preferably be rounded, otherwise the hook will bear upon two points only.

Latticed Poles.—Square, latticed, structural-steel poles may be of any width from that of true narrow-base poles used along curb lines to the wide poles which are in reality towers. There is no fixed dividing line between a pole and a tower, unless it be that of strength and rigidity, or possibly the use of widths which preclude shop riveting and shipment assembled. By far the

Fig. 57.—Steel crossing pole.

greater number of the structural-steel poles used are square in cross-section, one angle at each corner, and assembled and riveted before shipment. In the case of long poles, it will frequently be found advantageous to ship in two sections, and bolt them together in the field. There is no reasonable objection to the use of such field bolts, provided a splice is used of sufficient strength and length. The splice angle can be made an interior splice, with the root of the angle ground to fit the fillet of the main legs, and thus be comparatively unobtrusive in the final appearance of the pole.

Several types of poles are in use, the most common being

8

those with a regular bevel or those with parallel legs. Parabolic slopes have been used and they present a very graceful appearance under favorable conditions, although the rapid increase in width for longer poles may result in an inconvenient spread at the ground line. The parabolic slope has its true function in the application to very high towers of uniform height.

In some cases, the top portion of regularly sloped poles has been made with parallel sides in order to maintain like punch-

Fig. 58.—Railroad crossing pole, 13,000 volt wires.

ing, length and pin spacing of the crossarms. This does not always give a good appearance, however, on account of the marked bend at the lowest arm.

The foundation or anchorage of latticed poles may be made in three general ways, *i.e.*: by simply burying the lower portion, by providing separate ground-stub angles as in a rigid tower, or by a base plate or plates attached to anchor bolts. Further, the material entering the ground may be either painted or galvanized, although the former should be encased in concrete.

Base plates and anchor bolts are sometimes more expensive in material and workmanship than either of the other designs, but allow concrete foundations to be built in advance with the least probability of error in setting the connections to the super-structure.

Fig. 59.—Arrester house and guyed terminal pole.

As shown in Fig. 59, some quite elaborate bases have been used, the general purpose of these being to more firmly fix the base of the column and to provide an excess of material against cor-rosion. The real benefit derived from such construction, however, is open to question, and the gain does not seem com-mensurate with the cost. Steel poles can readily be protected against injury from accidental collisions, overflow, high water,

Fig. 60.—Two-circuit steel poles.

Fig. 61.—Steel poles with wooden arms, 11,000 volts.

etc., by the simple expedient of encasing the lower portion in concrete. Either a section of the latticed pole itself may be encased or long stub angles, embedded in high foundations, may be attached to the superstructure in the usual manner (Fig. 60).

Figs. 63 and 64 are views of the north and east sides respectively of a square latticed structural steel pole tested to

FIG. 62.

destruction. The views are unusually good in that they show very clearly the typical compression failure by buckling.

It will also be apparent that angle lacing is very effective in preventing buckling in the plane of the lacing, but that it has only a little restraining influence at right angles to its plane. It is further apparent that with latticed steel poles failure does not necessarily cause the pole to fall or break off. Instead the pole buckles near the base allowing the top to deflect, thus decreasing the wire tension but without causing the wires to fall to the ground.

The concrete foundations shown are relatively large for the poles they support, but were considered desirable in view of the location in a river bank.

The design of square latticed poles, in general, may be resolved into a determination of the stresses at the ground line or rather in the first panel above ground. This statement is based on

Fig. 63.   Fig. 64.
Latticed steel pole, test.

the assumption that, owing to the tops being relatively wider than in wood poles, the upper portion of the pole has an excess width as compared with the lowest panel. It is further predicated on there being no attempt made to seriously reduce the sections of the material in the upper half. In the case of parabolic slopes, stress determinations must be made at various heights since the widths presumably follow, more or less closely, the changes in bending moment, so the weakest section may be anywhere. Owing to the greater rigidity of pole frames, the breaking strength per unit of area in a pole will exceed that in a wide-base tower. Again, since the main legs have little inclination, the web system is compelled to carry the shearing stresses, which in a tower are partly carried by the main legs. For these reasons, the web or lattice is more often limited by the strength required than is the bracing of a wide-base tower. Shearing stresses must, therefore, be computed, and the lattice and its connection to the main legs designed accordingly.

Single flat lacing should not be used, except for small stresses and in narrow widths, since, as stated before, its strength is low and it is easily injured. Double flat lacing is applicable to greater stresses and widths, but it is often not as economical as angle lacing. In any case the strength of the pole depends on the unit strength of the weakest unsupported length, which is usually the lowest panel but may be the entire pole if the width is small and the height great. That is, the l/r of the entire cross-section of the pole may be greater than that of any single panel. The character and spacing of the lattice will determine to a large extent the amount of support afforded by it to the main leg angles at the panel points.

When both faces of a pole are connected with lacing at the same

FIG. 65.

level, the unsupported length of the main leg is the distance between panel points. If, however, the lacing is staggered, so that the support is in one direction only at each panel point, the unsupported length of the main leg is somewhere between a half and a whole panel length.

DESIGN OF STEEL POLE

One ⅜-in. Siemens-Martin galvanized stranded-steel ground wire.

Three No. 1 hard-drawn stranded copper, 33,000 volts.

Span　　= 400 ft.

Nor. sag　= 9 ft. 0 in.　Nor. tension (60°F.) = 570 lb.

Max. sag =10 ft. 6 in.　Max. tension (0°F., ½-in. ice, 8 lb. wind) = 1960 lb.

Fɪɢ. 66.—Design of steel pole.

Elastic limit, No. 1 wire = 2180 lb.

Wind pressure on wires (½-in. ice, 8 lb. per square foot wind):

　　⅜-in. ground wire = 0.917 × 400 = 367 lb.

　　No. 1 conductor　= 0.885 × 400 = 354 lb.

　　Wind on pole　　= 13 lb. per square foot × 1½ times exposed area

　　　　　　　　　　of windward side = 20 lb. per lineal foot,

With the above transverse loading and no broken wires, the compressive stress in the leg in the lowest full panel above the foundation is obtained by taking moments of the forces about the panel point 2 ft. 1 in. above the foundation.

$$
\begin{array}{lll}
\text{Ground wire} & 367 \text{ lb.} \times 42.4 & = 15{,}560 \text{ ft.-lb.} \\
\text{Power wires} & 354 \text{ lb.} \times 39.9 & = 14{,}120 \text{ ft.-lb.} \\
\text{Power wires } 354 \text{ lb.} \times 2 \times 37.4 & & = 26{,}480 \text{ ft.-lb.} \\
\text{Wind on pole} = 20 \text{ lb.} \times \dfrac{42.4^2}{2} & & = 17{,}980 \text{ ft.-lb.} \\
\hline
\text{Total bending moment} & & = 74{,}140 \text{ ft.-lb.}
\end{array}
$$

Since the lever arm of the resisting forces = 1.9 ft.:

$$74{,}140 \text{ ft.-lb.} \div (1.9 \text{ ft.} \times 2 \text{ legs}) = 19{,}500 \text{ lb.}$$

$$
\begin{array}{ll}
\text{Vertical load—steel} & = 1700 \text{ lb.} \\
\text{Vertical load—wires and insulators} & = 1500 \\
\hline
& 3200 \text{ lb.} \div 4 \text{ legs} = \quad 800 \text{ lb.} \\
\hline
\end{array}
$$

$$\text{Total compressive stress in each leg} = 20{,}300 \text{ lb.}$$

Since $1 \llcorner 3 \times 3 \times \frac{1}{4} = 1.44$ sq. in., the maximum unit stress in each leg $= 20{,}300 \div 1.44 = 14{,}100$ lb. per square inch.

If the side face of the pole is the same as the view shown, the leg is restrained from buckling in one direction by the intersecting diagonals, so the maximum l/r will be either the full panel length divided by the radius of gyration parallel to the leg, or the half panel length divided by the least radius of gyration,[1] *i.e.,*

$$\frac{l}{r} = \frac{44 \text{ in.}}{0.92} = 48 \quad \text{or} \quad \frac{l}{r} = \frac{22 \text{ in.}}{0.59} = 37$$

The ultimate strength of the angle based on the greater value of l/r and the curve in Fig. 33 is 35,000 lb. per square foot; therefore the factor of safety is $\dfrac{35{,}000}{14{,}100} = 2.5$.

Similarly the tensile stress in the other leg can be obtained by taking moments about the lowest panel point and subtracting the vertical load stress, which is always compression. For tensile stresses the area of the angle should be reduced by the area of one rivet hole.

Since the ultimate strength in tension is about 60,000 lb. per square inch, the compressive stress is generally the governing factor.

Stress in Diagonals.—The horizontal shear at the lowest crossarm is:

[1] See page 119, the assumption of the half panel length is conservative.

Ground wire   = 367 lb. × 1 =   367
Conductors   = 354 lb. × 3 = 1062
Wind on pole   = 20 lb. × 6 = 120

$$\text{Total} = 1549 \text{ lb.}$$

This will be carried by the web systems of two faces, making a shear of 775 lb. per face. With inclined legs, the stress in the diagonals will decrease from the lowest arm to the ground. Since the taper of the legs in this case is small, the stress in the diagonal just below the arm can be said, with only a small error, to equal the shear multiplied by the secant of the slope.

Assuming a 45° slope for the diagonals, the stress will be 775 lb. × 1.414 = 1100 lb.

Stress in Crossarms.—The crossarms should be designed for either the maximum ice and wind loads on both spans or for the maximum ice and wind loads on one span combined with a longitudinal load due to the breaking of the wire in the other span.

CONDITION No. 1

Vertical load (½-in. ice on wires) − 0.770 lb. per foot ×
            400 ft.................................... =   308 lb.
            Insulator and pin..................... =   25 lb.

$$\overline{333 \text{ lb.}}$$

333 lb. × 35 in....................... = 11,660 in.-lb.
Weight of arm = 15 lb. per foot × $\dfrac{3.1^2}{2}$ × 12 =   860 in.-lb.
Horizontal load (8 lb. per square foot wind on wire −
           0.885 lb. per foot × 400 ft.) =
           354 lb. × 12 in................ =   4,250 in.-lb.

$$\text{Total bending moment...... } = 16{,}770 \text{ in.-lb.}$$

2 ∟s 3½ in. × 3 in. × ⁵⁄₁₆ in. $\dfrac{I}{c}$ = 2 × 0.95 = 1.90.

Max. unit stress in crossarm = 16,770 ÷ 1.90 = 8800 lb. per square inch.

CONDITION No. 2

Vertical load (½-in. ice on wires) − 0.770 lb. per foot ×
            200 ft.......................... =   154 lb.
            Insulator and pin.................... =   25 lb.

$$\overline{179 \text{ lb.}}$$

$$179 \text{ lb.} \times 35 \text{ in} \ldots\ldots\ldots\ldots\ldots\ldots = 6270 \text{ in.-lb.}$$

$$\text{Weight of arm } = 15 \text{ lb. per foot} \times \frac{3.1^2}{2} \times 12 = 860 \text{ in.-lb.}$$

Horizontal load (8 lb. per square foot wind on wire —
$$0.885 \text{ lb. per foot} \times 200 \text{ ft.}) =$$
$$177 \text{ lb.} \times 12 \text{ in} \ldots\ldots\ldots\ldots = 2120 \text{ in.-lb.}$$

$$\text{Total bending moment} \ldots\ldots = 9250 \text{ in.-lb.}$$
$$9250 \div 1.90 = 4900 \text{ lb. per square inch.}$$

Longitudinal load:
$$1960 \text{ lb.} \times \frac{2.85}{1.50} = 3720 \text{ lb.} \div 1.94 \text{ sq. in.} = 1900 \text{ lb. per square inch.}$$

$$\text{Maximum unit stress in crossarm} \ldots\ldots = 6800 \text{ lb. per square inch.}$$

It should be noted that comparatively few insulators and pins can safely carry the longitudinal load assumed in Condition No. 2, even if the tie wires were able to transmit the load to the insulator. Generally, only strain poles or towers are able to

FIG. 67.—Curb-line poles.

fully meet this condition, as they are provided with either strain insulators or double-pin insulators and more effective tie or clamping devices.

Curb-line Poles.—Where high-voltage lines are located along curbs, it is important that the construction be of a high degree of excellence, and that structures be used which combine strength, a restricted width, and at least some esthetic qualities. A relatively high amount of insulation, with the consequent freedom from electrical failure, affords the greatest protection for the least expenditure. The width of poles must necessarily be restricted at the ground line, the maximum permissible width being from 24 in. to about 28 in. These dimensions are not fixed by any definite rule, but result from the precedent established by the use of large wooden poles. There are many places, however, where greater widths would not create any real obstruction nor presumably any active criticism. There should, in fact, be less objection to a line of well-designed steel poles than to wood poles, since their appearance is superior and only about one-half as many poles are required.

The surfaces of the poles which may possibly come into contact with pedestrians should be free from projecting edges; therefore, the latticing should be inside the main legs and its connections riveted rather than bolted.

It will sometimes be found advantageous to prevent the climbing of poles by unauthorized persons. This can be done by clamping wire netting against the lacing a short distance above the ground, or by filling the interior of the pole with concrete. The latter is not expensive, the forms being extremely simple, and it strengthens the pole both for general use and as a hub guard.

Triangular Poles.—The three-legged poles used heretofore have generally been of a proprietary type employing U-shaped main legs fastened at intervals with horizontal cast spreaders, but a few have been built of structural angles.

The material of the former is usually rerolled rail of greater unit strength but much harder and more brittle than structural steel. Owing to the shape and small flanges of U sections, as well as the hardness of the material composing them, it is not practicable to lattice the main legs by a true web system. In poles of this type the main legs are inclined more than is usual in square latticed steel poles, and the shearing stresses must be carried by the main leg sections.

In any structure having a triangular cross-section, the strength is not the same in all directions; therefore, three-legged poles are not well adapted by their form to withstand heavy loads.

When built of homogeneous material, which is difficult to insure in rerolled stock, such poles deflect considerably and will bend without actual fracture much more than square latticed poleṣ. Failure of three-legged poles should, therefore, be considered as occurring when a permanent bend is produced, or when the fastenings become loosened.

The logical service for poles of this design has been demonstrated by practice to be for supporting light lines in locations

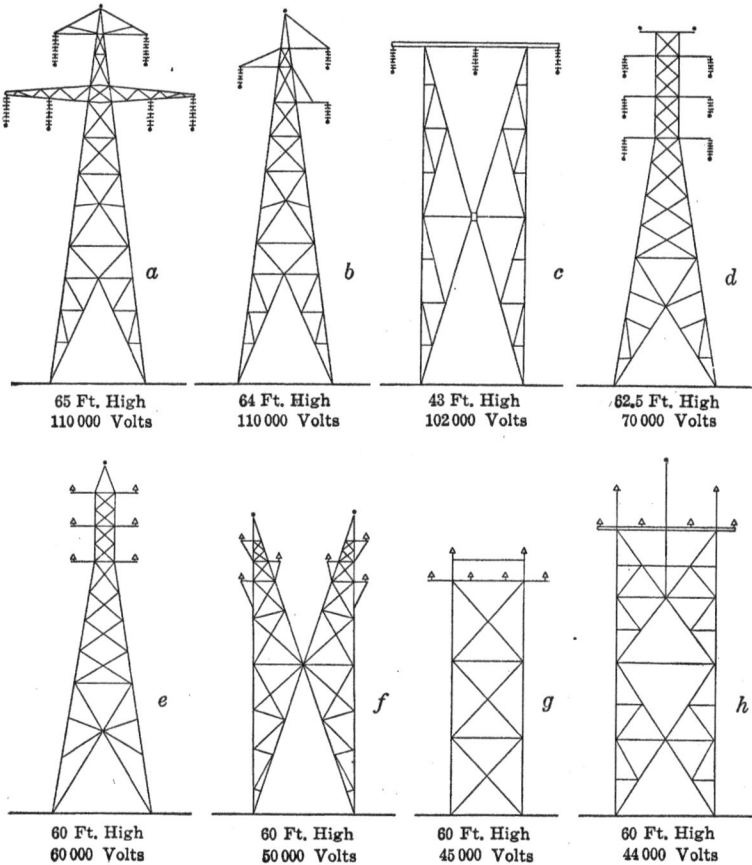

| 65 Ft. High | 64 Ft. High | 43 Ft. High | 62.5 Ft. High |
| 110 000 Volts | 110 000 Volts | 102 000 Volts | 70 000 Volts |

| 60 Ft. High | 60 Ft. High | 60 Ft. High | 60 Ft. High |
| 60 000 Volts | 50 000 Volts | 45 000 Volts | 44 000 Volts |

FIG. 68.—Types of towers.

involving difficult transportation, the poles being "knocked down," shipped in light-weight packages, and assembled in the field.

Wide-base Towers.—The forms of the frames which have been used in wide-base towers are of many types, as shown by Fig. 68*a*, *b*, *c*, *d*, *e*, *f*, *g* and *h*. The majority of designs are determinate frames, *i.e.*, those in which the stresses may be computed directly and definitely. Some, however, are what are termed indeterminate frames, since all the stresses cannot be computed directly owing to the fact that there are several paths through which the loads may be carried to the ground.

In such designs the actual distribution of stress will depend in part on the relative rigidity of the different paths. Although it is entirely possible to build indeterminate frames having any necessary strength, the practice involves a liability of error.

In general, the designs show a direct transfer of the tension and compression elements of the bending moment through the main legs, and a more or less complicated stiffening system whose chief function, in some cases, is to provide local support for the main legs.

The actual sections used for various members, even for somewhat similar installations, present marked variations. In fact there is no known structural theory which would make some designs desirable from either the purchaser's or the manufacturer's standpoint. For instance, a certain installation has 1 ∟ 4″ × 4″ × 3⁄4″ for the main legs, the panel length being 13 ft. The corresponding l/r is therefore 202, which is excessive for a main compression member.

In order to show more clearly the relative undesirability of the section in question, it may be compared with two other sizes of angle as follows:

	Area	Length	l/r	Breaking strength per square inch	Total breaking strength
1∟ 4 in. ×4 in. ×3⁄4 in.	5.44	13 ft.	202	12,000 lb.	65,200 lb.
1∟ 5 in. ×5 in. ×9⁄16 in.	5.31	13 ft.	159	16,000 lb.	85,000 lb.
1∟ 6 in. ×6 in. ×7⁄16 in.	5.06	13 ft.	131	20,000 lb.	101,200 lb.

It is evident from the foregoing that an ∟ 6″ × 6″ × 7⁄16″, having a much smaller value of l/r, would have been stiffer, stronger, lighter, and more readily fabricated than the section used. In fact, either the 5″ × 5″ or the 6″ × 6″ angle would have been a stronger and cheaper section.

It may be observed further that the actual factor of safety of the construction in question, under the maximum load assumed in its design, is a minus quantity. This is due solely to the excessive value of l/r or, in other words, inadequate bracing.

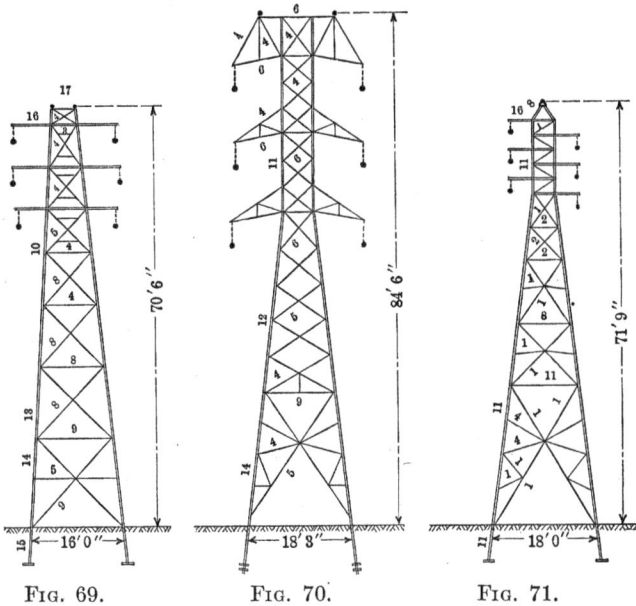

FIG. 69.   FIG. 70.   FIG. 71.

TABLE 27.—KEY TO TOWER SECTIONS. Figs. 69, 70, 71.

Mark	Section
1	L $1\frac{1}{2}$ × $1\frac{1}{2}$ × $\frac{1}{8}$
2	L $1\frac{3}{4}$ × $1\frac{3}{4}$ × $\frac{1}{8}$
3	L 2 × $1\frac{1}{2}$ × $\frac{1}{8}$
4	L 2 × 2 × $\frac{1}{8}$
5	L $2\frac{1}{4}$ × $2\frac{1}{4}$ × $\frac{1}{8}$
6	L $2\frac{1}{4}$ × $2\frac{1}{4}$ × $\frac{3}{16}$
7	L $2\frac{1}{2}$ × 2 × $\frac{1}{4}$
8	L $2\frac{1}{2}$ × $2\frac{1}{2}$ × $\frac{1}{8}$
9	L $2\frac{3}{4}$ × $2\frac{3}{4}$ × $\frac{1}{8}$
10	L 3 × 3 × $\frac{1}{4}$
11	L $3\frac{1}{2}$ × $3\frac{1}{2}$ × $\frac{1}{4}$
12	L $3\frac{1}{2}$ × $3\frac{1}{2}$ × $\frac{5}{16}$
13	L 4 × 4 × $\frac{1}{4}$
14	L 4 × 4 × $\frac{1}{4}$ × $\frac{5}{16}$
15	L 4 × 4 × $\frac{1}{4}$ × $\frac{3}{8}$
16	Chan. 4 in.—5.25 lb
17	Chan. 5 in.—6.5 lb.

Flexible Frames.—The steel A-frame trolley-wire support and transmission pole shown in Fig. 72 is typical of some of the lighter installations in Europe, where the A-frame was originated. This view is of additional interest in that the light brackets to which the trolley-wire messenger cables are attached would have little or no strength to withstand a broken cable, thus tending to show that failures in wires are not very seriously considered by the European designers of such structures. The illustration also shows a pair of grounding arms under the transmission circuits at the top of the pole.

TABLE 26.—RECORD OF SINGLE AND DOUBLE CIRCUIT WIDE BASE TOWERS. REPORTS (1915) FROM SIXTEEN LINES HAVING A TOTAL OF 7362 TOWERS IN SERVICE

Number of towers	Tower failures	Remarks
244	None	A crossarm twisted, due to burnt conductor.
378	None	
1041	None	
370	None	
64	None	
243	None	Three conductors burnt, due to contacts when heavy sleet unbalanced sags during removal.
1079	None	
593	None	
184	None	
33	None	
851	None	
110	None	
324	None	Four broken insulator connections.
913	None	A number of crossarm hanger rods have failed. Several conductors burnt.
748	1	Due to guy failure. Also about 70 breaks in conductors. (Storm of Apr. 2, 1915—worst on record.)
187	1	Compression leg of tower carrying 1250-ft. span. Also a number of slight buckles in other towers.
7362		(Storm of Apr. 2, 1915.)

When flexible frames were first used in this country, it was customary to insert a rigid or dead-ending tower at corners and at intervals of about five spans on tangents. In recent years, however, there has been a tendency to omit some of these stiffening structures on the theory that there was no real danger of the line falling longitudinally like a "house of cards." In view of a number of accidents that have occurred on such lines, it would seem desirable either to return to the former practice or to obtain the effect of stiffening structures by a more liberal use of guys.

A general objection to the flexible pole or frame is not intended, as they may have a proper usefulness in the construction of the lighter and less important lines. It is further probable that some criticisms of such construction would be more accurately directed to their shallow foundations, span lengths, details, and incorrect installation than to their use under favorable conditions.

Too much emphasis has been placed upon the need of flexibility, and to spend any efforts in providing greater flexibility than is found in the usual forms of support is a move in the wrong direction. A structure 40 ft. or more in height with only the resistance to bending inherent in such members acting as

Fig. 72.—A-frame trolley poles. European installation.

cantilever beams has naturally very much more flexibility than is required. To balance wire tensions only a slight movement of the pole top is required. Narrow-base A-frames have sometimes erroneously been termed poles or semi-flexible structures. A pole or tower is an enclosing or box-girder structure with four planes of bracing, whereas the narrow-base A-frame or latticed channel has but one central plane of bracing and is a true flexible frame.

Assuming that a reasonable amount of skill has been employed in selecting spans, heights and main sections, the next most important step in building an adequate A-frame line is to provide an overhead ground wire and substantial foundations. The ground wire, which should have considerable strength, may be given a little less sag than the conductors so it will serve as a continuous head guy, the value of which can hardly be overestimated. In fact, it is difficult to string the conductors unless there is a ground wire in place to steady the frames.

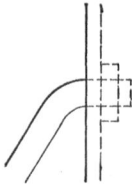

FIG. 73.

The foundations are also of great importance, since flexible frames are not well adapted to withstand eccentric loading. If the base of one main leg settles, or is erected at a different level than the other, the deviation of the top of the frame will be about seven times as much as the settlement, depending on the height and spread at the base. Moreover, as the failure of a frame will usually result from the buckling of the main channels, anything which disturbs an equal distribution of stress between the legs will promote failure.

That considerable foundation stress may be developed is shown by the fact that in a number of tests the bent rods used to attach the anchor members have been straightened out at the bend (Fig. 73). It seems probable, therefore, that more rigid attachments are needed.

FIG. 74.

The conditions which produce buckling are not very clearly understood, or rather their limits are not definitely known. If the main channels are assumed to be made of absolutely identical material and the base of the foundation is firm and unyielding, some difference in the lateral supports at the ground line or in the rigidity of the bracing connections may allow sufficient deflection to start buckling. As the failure is a compressive failure in a relatively long column, any measures which restrain the main legs from moving sideways at any point will be of effective service. A comparatively long stiff connection of the bracing with the main legs is useful as it stiffens the column locally. Such connections, therefore, should never be made with less than two rivets, and should preferably be not less than 6 in. long. Further, the diagonal braces should not have any slack, and if made of rods or other adjustable members, should be tightened as near equally as possible.

One of the most important requirements in A-frame construction is to provide guys at all corners above about 3°, and to pro-

Fig. 75.—A-frame failure.

vide strain towers at corners above 10°. This is in addition to a more or less definite number of head guys on tangents. In wire stringing, it is almost impossible to pull out the wires unless an overhead ground wire has been previously clamped in position to steady the frames. It is further necessary to pull all three wires of a circuit at one time, using a dynamometer and an equalizing rig to balance the tension in the wires. If an attempt is made to string one wire at a time, the frames may be twisted by the unbalanced loading. It is entirely possible to design and construct A-frame lines which will be satisfactory in cost and operation, but this cannot be done without the exercise of care and skill both in the design and in the erection.

The present tendency, and the writer ventures to believe it a proper tendency, is toward the use of galvanized ground-stub angles, whether the superstructure is painted or galvanized and with either concrete or earth back filling. Galvanizing such members is a relatively inexpensive operation. If desired, the galvanized surface can be painted over at the ground line. No reduction of section on account of the protective coating should be permitted in the ground stubs.

As flexible frames are painted as readily as narrow-base poles, the cost being in the neighborhood of $2 per structure, there is no objection as far as cost is concerned to the use of painted rather than galvanized superstructures.

RECORD OF FLEXIBLE A-FRAMES. REPORTS (1915) FROM SIX LINES HAVING A TOTAL OF 958 FRAMES IN SERVICE

Number of A-frames	A-frame failures	Remarks
158	None	
262	None	
62	None	Two cross arms twisted.
100	None	Four broken insulator connections.
290	3	Foundations pulled up.
86	56	Storm of Dec., 1914.
958		

TABLE 284.—TWO CIRCUIT TOWER LINES

No.	Name	Miles	Voltage	Insulators	Conductors Gage	Conductors Mat.	Ground wire No.	Ground wire Gage	Ground wire Mat.	Telephone Gage	Telephone Mat.	Two circuit towers Stand. span, ft.	Sag, 60°F.	Min. overhead clearance allowed	Wire spacing Arrangement	Horiz. ft.-in.	Vert. ft.-in.	R. of W.	Year built	Location	Design loadings Thickness of sleet (radial), in.	Wind on conductors	Wind on towers
1	Southern Indiana Power Co.	...	22,000	Pin	3	Cop.	1	...	Galv. steel	...	...	300	...	...	∴	3-0 diag.	...	...	...	Ind.	...	...	...
2	Penna. Utilities Co.	12	33,000	Pin	1	Cop.	1	3⁄8	Steel	None	None	500	...	...		...	...	...	1910	Pa.	¼ snow	26.0 lb.	26.0 lb.
3	U. S. Reclamation Service.	...	40,000	Pin	1	Cop.	...	...	None	...	...	400	...	...	∴	4-0 diag.	...	...	...		...	...	...
4	Santos Dock Co.	...	44,000	Pin		Cop.	1	5⁄16	Galv. steel	...	...	...	...	...	∴	...	...	...	...	Brazil	...	...	...
5	Utah Power & Lt. Co.	37	44,000	Susp.	0	Cop.	2	3⁄8	Steel	4 B.W.G.	Iron	600	16-0	...	···	6-0	6-0	...	1911	Utah	...	...	...
6	Utah Lt. & Rwy. Co.	14	44,000	Susp.	0000	Cop.	2	3⁄8	Galv. steel	...	...	500	7-4	...	···	14-0	5-0	...	1913	Utah	¼	9.0	
7	Arizona Power Co.	40	45,000	Susp.	4	Cop.	1	...	Galv. steel	...	...	...	...	30-0	···	10-0	5-0	...	...	Ariz.	...	...	...
8	Arizona Power Co.	35	45,000	Susp.	1	Cop.	1	...	Galv. steel	...	...	...	...	...	···	10-0	5-0	...	...	Ariz.	...	...	...
9	Penn. Central Lt. & Pwr. Co.	...	45,000	Pin	2	Cop.	...	...		...	...	500	...	...	∴	6-0 diag.	...	...	...	Pa.	...	...	...
10	U. S. Reclamation Service.	120	45,000	Susp.	00	Cop.	2	3⁄8	Galv. steel	...	...	800	...	...	···	8-0	6-0	P.	...	Ariz.	...	...	...
11	Canadian Lt. & Pwr. Co.	27	48,000	Pin	00	Cop.	1	...	Galv. steel	...	...	500	...	...		...	...	...	...	Canada	...	...	...

TABLE 28a.—Two Circuit Tower Lines. *Continued*

No.	Company																					
12	Portland Rwy. Lt. & Pwr. Co.	32	57,100	Susp.	250,000	Cop.	1	3/8	Galv. steel	10	Cop. clad	500	14-0	25-0	...	11-0	6-0	...		Ore.	1/2	10.6
13	Texas Power & Lt. Co.	204	60,000	Susp.	0	Cop.	1	3/8	Steel	Sep.	Sep.	650	13-6		...				1912 1914	Texas	1/4	50.0 mi.
14	Toronto Power Co.	80	60,000	Pin	198,000	Cop.	1	3/8	Galv. steel	10	Cop.	400	14-0	26-0	...	6-0 diag.				Canada	1/2	8.0
15	Western Canada Power Co.	32	60,000	Susp.	0	Cop.	1	3/8	Galv. steel			660			...							
16	Washington Water Power Co.	28	60,000	Susp.	270,000	Al.	2	3/8	Galv. steel	6 B.W.G.	Plough steel	650	21-0	29-3	...	15-2 26-6 19-10	7-0	100 P.	1910	Wash.	1/2	7.2 lb.
17	Great Northern Power Co.	16	60,000	Susp.	00	Cop.	1	5/16	Cop. clad			800	31-0		...	11-6	5-0	50 P.	1912	Wis.		
18	Great Northern Power Co.	15	60,000	Pin	00	Cop.	1	5/16	Galv. steel			400			...	6-0 diag.		100 P.	1906	Minn.		
19	Vancouver Power Co.	6½	60,000	Susp.	000	Cop.	1	3/8	Steel	None	None	600	14-2		...		7-0		1913	Canada	1/2	8.0
20	Carolina Power & Light Co.	26	66,000	Susp.	2	Cop.	1	3/8	Galv. steel			650	16-5				10-0	50 P.	1909	N. Car.		20.0
21	Central Georgia Power Co.	65	66,000	Pin	00	Cop.	1	4	Cop. clad			500	25-0		...	7-6 diag.		Ease-ment	1910	Ga.		
22	Central Georgia Trans. Co.	33	66,000	Susp.	000	Al.	2	3/8	Galv. steel			550	16-0	25-0	...			Ease-ment	1912	Ga.		
23	East Creek Lt. & Power Co.	29	66,000	Pin	2	Cop.	1	5/16	Galv. steel	6 B.W.G.		550	18-0	25-0	...	13-0	6-0	100 P.	1911	N. Y.	3/4	45.0 mi. actual
24	Northern Hydro-Elec. Co.	62	66,000	Susp.	0	Cop.	2		Cop. clad			520			...		6-0			Wis.		
25	Connecticut River Power Co.	66	66,000	Pin	2	Cop.	1	5	Cop. clad			410								Vt.		
26	Peninsular Power Co.	3½	66,000	Susp.	0	Al. Steel core	2	2	Cop. clad			540	18-0		...		6-4		1913	Mich.		20.0 2.0 area

TABLE 28a.—TWO CIRCUIT TOWER LINES. *Continued*

No.	Name	Miles	Voltage	Insulators	Conductors Gage	Conductors Mat.	Ground wire No.	Ground wire Gage	Ground wire Mat.	Telephone Gage	Telephone Mat.	Stand. span, ft.	Sag, 60°F.	Min. overhead clearance allowed	Arrangement	Horiz., ft.-in.	Vert., ft.-in.	R. of W.	Year built	Location	Thickness of sleet (radial), in.	Wind on conductors	Wind on towers
27	City of Winnepeg.	38	66,000	Pin	278,600	Al.	..	None	None	Sep.	Sep.	650	21-0	...		72-0	...	...	1909	Canada	½	8.0	25.0
28	Swedish S t a t e Rwys.	...	70,000	Susp.	40 sq. mm.	Cop.	1	35 sq. mm.	Iron	...	...	625	...	...		158	...	...	...	Sweden			
29	Penna. Water & Power Co.	40	70,000	Susp.	300,000	Al.	1	⅜	Galv. steel	Sep. poles	Sep. poles	500	17-3	25-0	:::	15-6	7-0	100P.	1910	Pa.-Md.	½	15.0	
30	Compania Hidro-Elec. e Irrig. del Chapala.	...	70,000	Susp.	99,000	Cop.	1	...	...	...	...	645	...	...		185	...	...	...	Mexico			
31	Southern California Edison Co.	66	75,000	Pin	0000	Cop.	..	None	None	...	...	700	...	...	.·.	6-0 diag.	...	...	...	Cal.			
32	Victoria Falls & Transvaal Power Co.	...	84,000	Susp.	60 sq. mm.	Cop.	3	35 sq. mm.	Galv. steel	...	...	500	...	...	.·.	9-2	...	...	...	Africa			
33	Toronto Power Co.	...	85,000	Pin	198,000	Cop.	1	⅜	Galv. steel	...	...	650	19-6	33-6	.·.	8-0	7-0	...	1914	Canada	½	8.0	
34	Mexican Lt. & Pwr. Co.	169	85,000	Pin	000	Cop.	1	...	Galv. steel	Sep. poles	Sep. poles	500	14-0	26-6	.·.	6-0 diag.	...	...	1910	Mexico	......	18.0	
35	Sao Paulo Elec. Co.	56	88,000	Pin	00	Cop.	1	⅜	Galv. steel	...	...	750	...	...		...	...	...	1913	Brazil			
36	Rio Janeiro Tram. Lt. & Pwr. Co.	51	88,000	Pin	000	Cop.	1	⅜	Galv. steel	10	B.B. galv. iron	400 500	...	...	.·.	8-0 diag.	...	...	1913	Brazil			

TABLE 28a.—Two Circuit Tower Lines. *Continued*

No.																				
37	Southern Power Co. 241	88,000	Susp.	00 0000	Cop. Al.	1	⅜				600			16-0	8-0	P.	1909 1913	N. Car. S. Car.		
38	Southern Power Co. 175	88,000	Susp.	00 0000	Cop. Al.	1	⅜	Galv. steel			600		25-0	18-0	10-0	P.	1911 1913	N. Car. N. Car.		
39	Shawinigan Water & Pwr. Co. 87	100,000	Susp.	250,000	Al.	2	⅜	Galv. steel			520				8-0		1911	Canada		
40	Los Angeles Aqueduct 47	100,000	Susp.	250,000 300,000	Cop.	1		Galv. steel			650				10-0			Cal.		
41	Tata Hydro Elec. Co. 43	100,000	Susp.		Cop.	1					500			10-6 diag.			1914	India		
42	Great Western Power Co. 154	100,000	Susp.	000	Cop.	1	⁵⁄₁₆	Galv. steel	8	Cop.	750	15-0	35-0	14-0	10-0	70 P.	1908	Cal.		
43	Yadkin River Power Co. 96	100,000	Susp.	1	Cop.	1	⅜	Galv. steel	Sep. poles	Sep. poles	650	13-6	30-0	15-0	9-0	50 P.	1911	N. Car.	¼	50.0 mi.
44	Sierra & San Francisco Pwr. Co. 138	104,000	Susp.	00	Cop.			None	Sep. poles	Sep. poles	800	16-6	30-0	15-0	9-0	Ease-ment	1910	Cal.	10.0 lb.	10.0 lb.
45	Hydro-Elec. Pwr. Commission. 280	110,000	Susp.	00 0000	Al.	2 3	⁵⁄₁₆	Galv. steel			550	14-0	28-0		10-0	Ease-ment	1910	Canada		
46	Mississippi River Power Co. 144	110,000	Susp.	300,000	Cop.	1	½	Galv. steel	Sep. poles	Sep. poles	800	18-8	30-0	17-11 18-7 19-11	10-0	100 P.	1912	Iowa	½	6.0 lb. 30.0 lb.
47	Georgia Rwy. & Power Co.	110,000	Susp.	0000	Cop.	2	⁷⁄₁₆	Galv. steel			550							Ga.		
48	Alabama Power Co.	110,000	Susp.	00	Cop.	2		Galv. steel			600 700	17-0		10′-0 diag.				Ala.		
49	Lehigh Navigation Elec. Co. 36	110,000	Susp.	250,000	Cop.	2	⅜	Galv. steel			600			10-0		P.	1913	Pa.	½	8.0
50	Cedar Rapids Mfg. & Pwr. Co.	110,000	Susp.	500,000	Al.	1					660				10-0		1913	N. Y.		8.0 lb. 1.5 area

TABLE 28a.—TWO CIRCUIT TOWER LINES. *Continued*

No.	Name	Two circuit towers																					
		Miles	Voltage	Insulators	Conductors		Ground wire			Telephone		Stand. span, ft.	Sag, 60°F.	Min. overhead clearance allowed	Wire spacing			R. of W.	Year built	Location	Design loadings		
					Gage	Mat.	No.	Gage	Mat.	Gage	Mat.				Arrangement	Horiz., ft.-in.	Vert., ft.-in.				Thickness of sleet (radial), in.	Wind on conductors	Wind on towers
51	Mexican Northern Pwr. Co.	...	110,000	Susp.	....	Al.	3	3/8	Galv. steel	....	....	575	......	....	.: :.	10–0 diag.	....	.....	1913	Mexico			
52	Ebro Irrigation & Pwr. Co.	105	110,000	Susp.	0000	Cop.	1	3/8	Galv. steel	....	....	750	......	....		......	8–0	.....	....	Spain			
53	Pacific Gas & Elec. Co.	140	110,000	Susp.	000	Cop.	..	.....	None	None	None	800	23–0	....	: :	......	10–0	.....	1913	Cal.	3/4 ice +14 snow	14.5	
54	Inawashiro Hydro-Elec. Pwr. Co.	144	115,000	Susp.	100 sq. mm.	Cop.	2	3/8	Galv. steel	....	....	550	......	....		......	10–0	.....	1913	Japan			
55	Connecticut River Trans. Co.	60	120,000	Susp.	00	Cop.	1	3/8	Galv. steel	....	....	600	......	75–0	: :	......	10–0	.....	....	Mass.			
56	Tennessee Power Co.	...	120,000	Susp.	00	Cop.	1	5/16	Galv. steel	6	Cop. clad	663	......	....		10–6		.....	....	Tenn.			
57	West Penn. Trac. & Water Pwr. Co.	106	125,000	Susp.	0	Cop.	2	4	Cop. clad	....	....	528	......	....	: :	......	5–0	.....	1913	Pa.			
58	Utah Power & Lt. Co.	135	130,000	Susp.	250,000	Cop.	2	3/8	Galv. steel	None	None	650	12–9	....	: :	......	13–0	.....	1913	Utah	1/4	10.5 lb.	65.0 mi.
59	Southern Sierras Power Co.	239	140,000	Susp.	....	Al. steel core	1	3/8	Galv. steel	Sep. poles	Sep. poles	660	13–0	30–0	.: :.	16–6 26–6 16–6	10–0	.....	....	Cal.	1/2	11.7	

TABLE 28b.—ONE CIRCUIT TOWER LINES

One circuit towers

No.	Name	Miles	Voltage	Insulators	Conductors Gage	Conductors Mat.	Ground wire No.	Ground wire Gage	Ground wire Mat.	Telephone Gage	Telephone Mat.	Stand. span, ft.	Sag, 60°F.	Min. overhead clearance allowed	Wire spacing Arrangement	Horiz. ft.-in.	Vert. ft.-in.	R. of W.	Year built	Location	Thickness of sleet (radial), in.	Wind on conductors	Wind on towers
1	Utah Power & Lt. Co.	37	44,000	Pin	115,570	Cop.			None	None	None	800	19.9			12-0			1909	Utah			
2	Utah Power & Lt. Co.	10	44,000	Pin	4	Cop.			None	None	None	350			·.·	6-0	6-0		1906	Utah			
3	La Crosse Water Power Co.		46,000	Pin	2	Cop.	1	2	Cop.	5	Cop.	480			·.·	6-0 diag.							
4	Niagara, Lockport & Ontario P. Co.		60,000	Pin	428,000	Al.						550									½	15.0	
5	Adirondack Elec. Power Corp.	18	60,000	Pin	000	Cop.	1	000	Cop.	12	Cop. (mess)	550	18-0		·.·	6-0	6-0		1906	N. Y.			22.5
6	Car. Pwr. & Lt. Co.	26	66,000	Susp.	2	Cop.	1	⅜	Galv. steel			700	17-0			10-4¾		50P.	1909	N. Car.	3	50.0 mi.	
7	Telluride Power Co.		88,000	Pin	0	Cop.			None			900				12-0			1909				
8	Colorada Power Co.	183	100,000	Susp.	0	Cop.			None			750				10-0		100P.	1909	Col.			

TABLE 28b.—ONE CIRCUIT TOWER LINES. *Continued*

	Company		Voltage				Material			Ground wire		Span							Year	Location	Ice/snow		
9	Great Falls Power Co.	132	102,000	Susp.	0-	Cop.	2	3/8	Galv. steel			600							1910	Mont.			
10	Yadkin River Power Co.	21	104,000	Susp.	1	Cop.	1	3/8	Galv. steel			650	16-2		15-0	9-0	50P.	1912	N. Car.			
11	Sierra & San Francisco Pwr. Co.	98	104,000	Susp.	00	Cop.						800	18-0	30-0	15-0	9-0	Easement	1909	Cal.			
12	Av. Sable Elec. Co.	35	110,000	Susp.	2	Cop.			None			528		53-0 60-0	8-0	d ag		1906	Mich.			
13	Chile Exploration Co.	85	110,000	Susp.	000	Cop.	2	3/8	Galv. steel			650	19-6 55°C.)		12-4			1914	Chile		18.0 lb.	30.0 lb.
14	Pacific Gas & Elec. Co.	12	110,000	Susp.	000	Cop.			None	None	None	400 500	10-0		13-6			1913	Cal.	3/4 ice +1 1/4 snow	14.5 lb.	14.5 lb.
15	Columbus Power Co.	60	114,000	Susp.	00	Cop.	1	3/8	Steel	10	Cop. clad	600	10-0			10-0		1912	Ala.-Ga.			
16	Eastern Michigan Power Co.	245	140,000	Susp.	0	Cop.			None			500	12-0	25-0	12-0	12-0	4-10 rods	1911	Mich.			
17	Pacific Lt. & Power Co.	241	150,000	Susp.		Al.	1	1/2	Galv. steel			660		48-0	17-6			1913	Cal.			

TABLE 28c.—ONE CIRCUIT STEEL POLES

Name	Miles	Voltage	Insulators	Conductors		Ground wire			Telephone		Stand. span, ft.	Wire spacing			Year	Location
				Gage	Mat.	No.	Gage	Mat.	Gage	Mat.		Arrangement	Hor.	Vert., ft.		
Societa Italiana di Elettrochimica.	.....	88,000	Pin	66 sq. mm.	Cop.	.....	.....	.....	.....	.....	656		7-1	.....	.....	Italy
Societa Generale Elellrica dell'Adamello.	.....	72,000	Pin	79 sq. mm.	Cop.	.....	.....	.....	.....	.....	607	∴	6-3	.....	.....	Italy
Societa Elettrica Riviera di Ponente.	.....	70,000	Pin	78	Cop.	.....	.....	.....	.....	.....	494	∴	5-5	.....	.....	Italy
Muncie Electric Light Co.	.....	60,000	Susp.	1	Cop.	.....	.....	.....	.....	.....	.....	∴	6-2	6-6	1912	Indiana

TABLE 28d.—TWO CIRCUIT STEEL POLES

Name	Miles	Voltage	Insulators	Conductors		Ground wire			Telephone		Stand. span, ft.	Wire spacing			Year	Location
				Gage	Mat.	No.	Gage	Mat.	Gage	Mat.		Arrangement	Hor.	Vert., ft.		
Sanitary District, Chicago.	30	44,000	Pin	.....	Al.	.....	.....	None	.....	.....	350	∴∵	6-0	.....	.....	Illinois
Lauchhammer, A. G.	35	110,000	Susp.	1	Cop.	1	100,000	Steel	.....	.....	550	∵∴	5-10	.....	.....	Germany
North. Co., G. & E. Co.	15	22,000	Pin	1	Cop.	1	.....	Steel			450	∴			1914	Pa.
Central Georgia.	2	66,000	Susp.	000	Al.	2	⅜	Steel			425	∵∴			.....	Georgia
Southern California Edison Co.	7	60,000	Susp.	0000	Cop.	.....	.....	None			660				.....	Cal.
Southern California Edison Co.	2	60,000	Susp.	0000	Cop.	.....	.....	None			200				.....	Cal.
Amherst Power Co.	½	66,000	Susp.	0	Cop.	1	.....	.....			300	∴∵	13-0 16-0	9-0	1913	Mass.
Manchester Tract., L.&P.Co	.....	33,000	Pin	2	Cop.	1	.....	Steel			300	∴∵	3-0	2-8	1914	New Hamp.

CHAPTER VII

SPECIAL STRUCTURES

Transposition.—In addition to the regular line structures, provision must be made for supports on which the conductors may be transposed. Transposition poles or towers are few in number and their divergence from the standard type may or may not be marked. In some cases it is only necessary to change the attachment of the conductors to a different one of the usual

Fig. 76.—Wooden pole terminal rack.

points of connection. In other cases structures with greater conductor spacing or specially arranged crossarms, etc., may be required. Under any circumstances, whether the normal separation be small or great, care must be taken not to unduly diminish the clearance between conductors where they cross in the span.

141

Structurally, transposition consists of interchanging the points of conductor attachment, so that no one conductor maintains its original location throughout the entire line. The number of transpositions and the distances between the points at which they are made vary greatly for different installations, voltages, etc. It is possible that the introduction of special transposition struc-

Fig. 77.—Outdoor sub-station.

tures might be reduced by the combination of the transposition towers with special towers otherwise required at various points. A more frequent form of transposition is that customary in the telephone wires carried on transmission line supports. This may be done at every few poles on short-span lines or in every span on long-span lines. Owing to the smaller clearances required between such wires, however, the measures for their transposition will frequently consist merely of providing two points about a foot apart vertically for the attachment of the arm or bracket angle.

Another method is to cross the wires inside the tower, by using pins of different height or placing one pin on a raised shelf angle.

Outdoor Sub-stations.—Outdoor sub-stations in which the apparatus is mounted on poles, towers, or frames, instead of being inclosed in a building, have come into considerable use in the last few years, although the propriety of their use under adverse climatic conditions has been the subject of criticism. Such stations are of various types, depending on the voltage, location, and importance of the line. In the majority of instances, the transformers have been elevated and supported on platforms from 10 to 15 ft. above the ground. By so doing, the transformers are removed from the zone of possible contact with passersby, but at the expense of accessibility and with the addition of heavier and more costly supports. When the transformers are placed at the ground level, it is frequently desirable to inclose them in metal or wire screens in order to isolate them from contact.

Switching Stations, Etc.—Outdoor switching stations, transformer stations, etc., have come into very general use in some form or another in recent years. In the simplest type they consist of special arms, platforms, etc., on one or more wood poles, involving no particular changes from the standard line construction. The more elaborate designs are dead-ending frames, of timber or steel, with special supports for insulators, switches, arresters, etc., and platforms or housing for transformers.

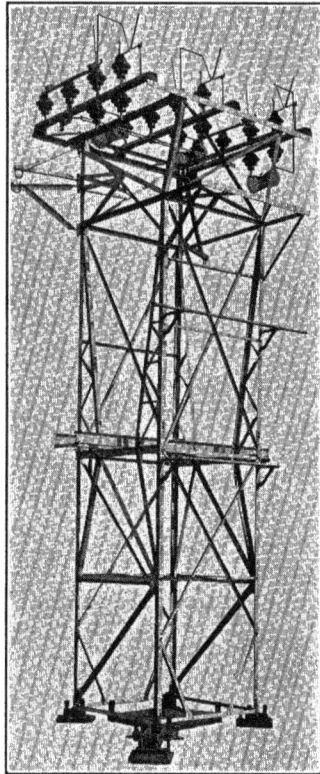

Fig. 78.—Outdoor sub-station.

The very large clearances necessary with high voltages have been the cause of a number of rather pretentious structural frames.　In form, these towers or series of connected towers, as the case may be, have not adhered to any particular outline, their shape usually depending on local conditions and requirements. Inasmuch as such structures usually serve as dead-end supports, or worse, as sharp-corner towers, and have a large area exposed to the wind, they should be relatively much stronger than the

Fig. 79.—Single-circuit tower, flat spacing (emergency switches).

standard line supports.　There has unquestionably been some tendency toward the use of excessive length ratios, and of inconsistent connections at the foundation.　It must be remembered that a structure which terminates at the ground line in a single angle at each corner requires those angles to transmit all the shear in addition to their tensile or compressive stresses.　Such construction frequently appears, and in fact is, top heavy.

High Towers.—Unusually high towers are required at river cross-ings, or where one or both ends of a very long span are on low

Fig. 80.—Switching frame.

Fig. 81.

ground. In the former case permission to make the crossing, if over a navigable stream, must be obtained from the War De-partment. The requirements, heretofore, have been merely that

10

there should be no encroachment on government lines, and that a certain minimum overhead clearance be provided. The clearance varies with the importance of the stream, ranging from 120 ft. over creeks or small rivers to 150 ft. over deep-water highways. The structures for such crossings may be of three types: guyed masts, semi-dead-end towers, and dead-end towers.

The desirability of any one of these, for a particular crossing, will depend on the number of wires, the ground space available,

Fig. 82.—Guyed masts 151 ft. high, 28 wires.

and the importance of the line. If the character of the adjoining supports is such that they may be depended on to maintain a stretch of wires of which a reasonable percentage is unbroken, then there is no need of the high structures being "self-supporting." If only a limited ground space can be obtained recourse must be had to guyed masts, the strength of which lies in the unbroken wires and the guys. Towers capable of completely dead-ending a small number of wires are entirely feasible, but to dead-end heavy lines with no allowance for pullback or guys would require considerable expense and massive structures.

The towers in Figs. 82, 83, and 84 support probably the heaviest lines thus far carried in overhead river crossings. None of these structures was designed to dead-end the full number of wires under ½ in. ice and 8.0 lb. wind load. They were, however, designed to withstand the maximum transverse loading, combined with some unbalancing, such as a few broken wires, with a factor of safety. Since the normal wire tension is approximately only one-third of the maximum, they would also presumably withstand normal dead-ending. Some additional security is obtained by longitudinal guys away from the river, since failure of many wires could result only from accident to adjoining structures. It may

FIG. 83.—Semi-dead-end towers.

be noted that there is no grading up of the adjoining supports, the rise of the wires to the high towers being made in one span. This method was first adopted as a standard construction on these installations, and the results have been entirely satisfactory. Since the change in elevation of the wires is from a height of about 50 ft. to one of 150 ft., it is evident that a number of high approach towers are avoided. Such an abrupt inclination of the wires is unsuited to pin insulators because the wires would not clear the petticoats in descending and would pull the insulators from the pins in ascending, except that a saddle covering a double series of pin insulators might be used. The most satisfactory arrangement, and the strongest mechanically, is to use strain insulators of the types shown in Figs. 115 and 117.

On the high towers the insulators may be in the strain position or they may be suspended. On the adjoining supports they may be in the strain position or in the suspended position *under* the wire. In the latter case the pole or tower next to the river tower is generally subjected to uplift and has no downward load.

Care must be exercised in the design and in the wire stringing to balance the tension in the crossing and adjoining spans, other-

Fig. 84.—Semi-dead-end towers.

wise the high towers will be subjected to a heavy loading and the underhung insulators noticeably deflected from the vertical. When the crossing span is much longer than the approach spans, as is frequently the case, the wires in the inclined span will have very small sags and may be brought together on the low pole to the standard spacing for the voltage in question, even though a considerable separation is used to prevent swinging contacts in the long span.

All high towers should be provided with overhead ground wires and ladders, and should be grounded below the foundations.

As these structures are relatively heavy it is more economical to paint them than to use galvanized material, the saving in original cost being more than ample to maintain a high degree of protection with paint.

Aerial Cable.—In some instances when an overhead line has had to cross territory subject to restrictive requirements, and particularly when adjacent to submarine or duct lines, aerial insulated cables have been used. This arrangement is practicable

Fig. 85.   Fig. 86.
Aerial cable.

only for voltages below about 22,000, because satisfactory insulated cable is not obtainable for higher voltages.

In the installation shown in Figs. 85 and 86, the cables are hung from steel messengers and supported on a low line of closely spaced poles. The clearances necessary are those due to the physical requirements for passageway beneath the cables. Short heavy poles are used, being guyed at corners and ends but with no attempt to design for broken messengers or cables.

In the background of Fig. 86 may be seen a heavy H-frame, telegraph and telephone line.[1]

An aerial duct line for 13,000-volt, three-wire, insulated cables, attached to the side of a railroad bridge, is shown in Fig. 87. Provision was made for three circuits and one spare. As shown,

FIG. 87.—Aerial duct line.

the cables are encased in split metal tubes supported on steel brackets, riveted to the outer stiffeners of the bridge. This construction was considered advisable as the line had to cross two railroads, one heavy high-voltage line on tall towers, and several very important telephone and telegraph systems, and

[1] Subsequently wrecked by a very severe sleet-storm.

there was not enough space available between the existing power and telegraph systems.

In making such attachments to bridges, and to some extent to any foreign structures, the construction must be accessible without interference with other interests, and at the same time be free from the probability of injury by foreign workmen.

CHAPTER VIII

CONCRETE POLES

Wood poles had and still have certain advantages; they also have certain disadvantages. Some of the good points cannot be duplicated in concrete, but on the other hand, some of the objectionable features can be eliminated. Therefore, omitting undue enthusiasm on the one hand and any pretensions to magic excellence on the other, the matter is purely and simply one of final cost and final efficiency.

It is self evident that concrete poles can be made, and it is fairly well known that a few thousand have been made and are now in service. The only remaining considerations seem to be ones of mechanical efficiency and actual or proper cost. The question of mechanical efficiency is in reality combined with that of cost. Given time and money enough there are few structures which cannot be constructed, even by an amateur. Economical construction is another matter, and by economical is meant true economy—final economy—not merely reduced first cost. Thus, in the case of pole construction, we find that some have been built at a low initial cost but with an equally low mechanical efficiency, while others have been built at abnormal expense and with excessive strength. Neither extreme is good engineering nor good economics.

If an important wood line is to be built, stronger poles are used than for an unimportant line. If the species of timber has great strength and the poles are "selected," smaller poles may be used than if the timber is sappy, knotted, etc. Likewise, if concrete poles are to be used, the importance of the work and the dependability of the concrete should influence the specified size and strength. Further, and in view of the very rare mechanical failures of undecayed timber poles, it seems essential to consider the actual strength of the poles which are and have been carrying various installations of wires. The strength may have been too great—it was presumably not too small—but it serves in some degree as a measure of the desirable strength for concrete poles which may properly replace timber ones.

152

This method neglects the relative factors of safety, reference to which will occur elsewhere, but it is a fact that those who will without hesitation use any mean, crooked little timber pole will develop a multitude of rules and requirements when building a concrete pole.

FIG. 88.—Hollow concrete poles. European transmission line.

The writer does not wish, in any degree, to convey approval of weak or shoddy construction, but would emphasize the need for intelligent analysis of the subject. Whatever the future of concrete poles, it can be of no service to the power or telegraph business to construct either weak poles or ridiculously strong ones. It cannot be denied that individual success in the concrete-pole industry will depend very largely on the degree in which the fabricator is successful in maintaining a uniform product.

In the case of wood poles, there has always been a wide range in the strength even within the same species, greater differences than are perhaps generally realized. These differences together with decay have been guarded against by large safety factors—or perhaps more accurately by the use of apparently large factors— the factors being actually less than supposed. In many cases there has been the utterly indefensible practice of assuming impracticable loadings together with impossible unit stresses. There can be no engineering justification for claiming that poles are designed to withstand certain improbable loads with a factor of safety of 5 or 6 when, in fact, 99 per cent. of the poles would not be subjected to such loads, nor could they carry anything like five times such loads.

When steel is embedded in well-made concrete its preservation is perfect and the life of a reinforced monolith is practically indefinite. If designed and built with the same attention now given other materials, reinforced-concrete poles should attain the necessary strength and give satisfactory service. Like steel poles, they can be spaced greater distances apart than is economically possible with wood poles, and their fire-resisting qualities are about equal to steel poles. This latter feature will become of increased importance with the spread of modern requirements for fire protection.

In damp climates, or in localities where wood poles are subject to attack by fungi, or insects, concrete poles have a longer life than either steel or timber. By the insertion of pipes, or the formation of an axial passage in the concrete, wires may be carried from the pole tops to the ground and thence in any desired direction and are thus entirely protected at little additional cost.

Owing to the natural taper of timber, it is frequently the case that the weakest section is at some point above the ground level. Therefore, there is an excess of material in the butt, which may be considered wasted, except in so far as this surplus timber is useful in resisting decay. A reinforced-concrete pole may be given any desired taper and need have little or no excess of improperly placed material.

If we may judge by the kind of handling which concrete piles successfully withstand, it seems entirely probable that concrete poles, if properly reinforced, will survive any shocks incident to ordinary service. When subjected to an overload or accidental shock, a timber pole will bend and in some cases survive; but failure when it does occur is usually complete and the pole falls. Concrete poles, on the contrary, while without the elasticity of timber, do not fail by breaking off, but are held by the reinforcement from falling to the ground. Tests also

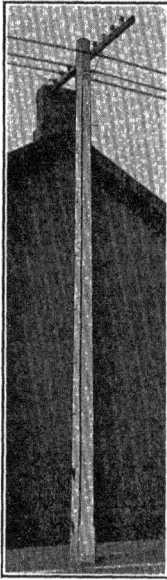

Fig. 89.   Fig. 90.

Concrete poles.

show that a reasonable amount of bending (sufficient to balance stresses in the wires) can occur without apparent injury to the pole.

The chief cause of skepticism heretofore has been the fear that such long, slender concrete beams would not be able to withstand, without cracking, the bending stresses and measurable deflections of a pole line. If the poles are properly designed, cracks due to partial failure will not occur. Hair cracks are of infrequent occurrence, microscopic in character, and experience has apparently shown that they will not admit moisture in sufficient quantity to injure a reinforced-concrete structure.

Considerable misinformation has been published on the subject of concrete poles. Generally this has taken the form of cost data, which could not be duplicated in actual practice. The plain fact of the matter is that something cannot be obtained for nothing even in concrete. If a concrete pole of considerable strength is required, the material needed to provide that strength must be purchased and placed in the pole. No costs are directly comparable unless they relate to poles of similar strength and similar quantities. For instance, 30-ft. poles having an ultimate strength of about 1000 lb. have been made for approximately $8, but a 30-ft. pole having a strength of 6000 lb. would require reinforcement the cost of which alone would exceed the entire cost of the weaker poles.

The number of poles required has a very direct bearing on the cost and on the proper place of manufacture. The cost of forms, plant, superintendence and engineering must be borne proportionally by the number of poles to be made by that manufacturing plant. Again, the plant and forms for a small number of poles would not be ultimately economical for a large installation. Too little attention has been paid to the very real cost in time and money required to properly develop the engineering side of the question. It is not beyond the bounds of reason to believe that the costs of investigation and engineering if included in some published data would have more than doubled the stated costs.

One criticism which can properly be made of concrete poles is that they have considerable weight. This affects both the cost of handling and the amount of the freight, if the poles are made in a distant yard. In an attempt to reduce this cost, poles have been cast in place and the claim made that an economy is effected thereby. Such procedure is open to serious criticism, both on the score of the probable excellence of the final work and as to the actuality of the economy claimed. It would seem difficult to make any experienced constructor believe in the economy of moving the forms, water and aggregates from pole to pole and casting in the upright position with the delays consequent to such an arrangement.

A number of what may be termed freak designs have been given more or less extended trial, but they appear to have little to commend them. Among these may be mentioned the combination wood and concrete pole consisting of a wooden pole incased in a shell of reinforced concrete, the triangular or un-climbable pole, the sectional pole and the type with hollow-paneled sides.

Concrete-incased timber has no claim to efficiency, either theoretically or practically and it is not believed that the type will ever be given another trial. Triangular concrete poles possess the disadvantage common to all triangular poles—that

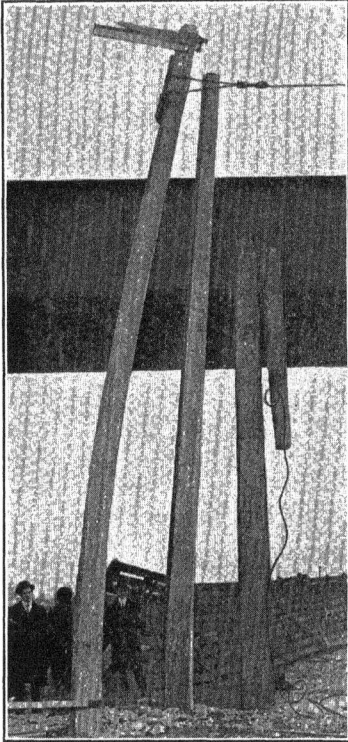

FIG. 91.—Coombs concrete pole
(test pull 6000 lbs.)

FIG. 92.—Coombs concrete pole
(test pull 7200 lbs.)

of having much less strength in one direction than in the opposite direction. By the addition of the small amount of material needed to form a square, either a stronger pole is obtained, or a reduction in outline or reinforcement made possible.

Hollow-sided poles have been used to some extent in Germany, but the only advantage which would appear to justify their use here is reduced weight, and this can be secured by using a hollow-axis pole. Hollow panels or steps reduce the strength in one direction and apparently impose a heavy arch duty on the steps to care for shear and buckling stresses. Furthermore, it is doubtful whether such poles situated in a locality subject to snow and sleet would be climbable.

It has been claimed by some that if cracks develop in concrete poles they may be readily filled with cement, the inference being that the pole is then restored to the condition in which it should have been originally. There is some basis of fact in this contention, if the cracks in question are seasoning cracks. If, however, the pole develops tension cracks due to faulty design, plastering with cement grout will not prevent a repetition of the cracking on a subsequent application of the load. In addition, the practical application of a plastering process to poles in service contemplates a rigidity of inspection and cost of maintenance entirely at variance with the economic theory of the use of concrete poles. In fact this question of cracks is one source of reasonable criticism of concrete poles, as the possible corrosion of the reinforcing metal would be a very serious matter. Thus it has been pointed out that owing to the thin shell of concrete outside the rods— usually about 1 in. thick—there may be absorption of moisture even when no hair cracks or deflection cracks are present.

In some locations, notably that of the Pennsylvania Railroad poles designed by the writer for the line on the New Jersey meadows, the constantly wet foundation and the size of the line —this being the heaviest concrete installation in the country— renders this possibility of importance in the design. The soil in question is a peaty bog and always holds water. Above ground there is usually water during portions of the year although it is fairly dry in summer. In designing the above line, it was felt that by making the poles of considerable strength, providing for local stresses and manufacturing under competent supervision, the probability of cracks or corrosion would be very nearly eliminated.

Since in solid poles of light capacity the loading produces a low compressive unit stress in the concrete, a considerable area of concrete might be omitted, or, theoretically, the economical section would be a hollow one. The relatively greater weight of a solid pole makes it more difficult to handle; therefore, a hollow pole would be more economical to erect. Further, the sides of the pole which resist bending stresses normal to the line might be at a greater distance from the center than the sides perpendicular to the line.

There are, however, certain objections to the use of hollow or unsymmetrical sections. The former are difficult to make properly and the cost of the forms exceeds that required for solid sections. Unsymmetrical sections may perhaps be open to criticism on the score of appearance and if the lack of symmetry is very pronounced, the poles will be relatively weak in one direction.

In general, a square, octagonal, circular or other cross-section may be used, but as a matter of appearance it is desirable that all corners be chamfered or rounded since sharp edges are difficult to make and easily broken. The minimum diameter, or width, at the top may be made 6 in. for small poles and increased as required for the strength and appearance of long poles or poles carrying a heavy line. In any case care should be exercised, in determining the taper and reinforcement, that no weak section occurs at some distance above the ground-level.

In regard to the factors of safety, unit stresses and working stresses, to be allowed in the constituent materials of a reinforced-concrete pole, there is as much latitude of judgment as in other structural work. The character of

FIG. 93.   FIG. 94.

FIG. 93.—Torsion test on pole not provided with spiral reinforcement—diagonal crack near top.

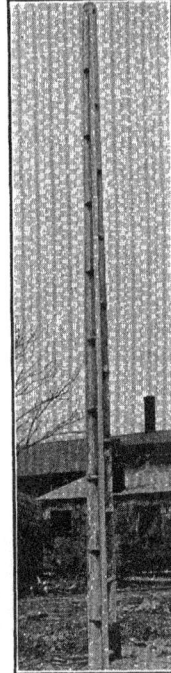

FIG. 94.—European + shaped pole.

service is not closely akin to that of bridges or buildings, so the factors of safety common to such work would be too conservative for poles computed for extreme conditions of loading.

The present practice differs rather widely as to the most economical or most desirable distribution of reinforcement, but, it is now generally conceded in reinforced-concrete work that the finer the distribution of metal, the greater will be the homogeneity

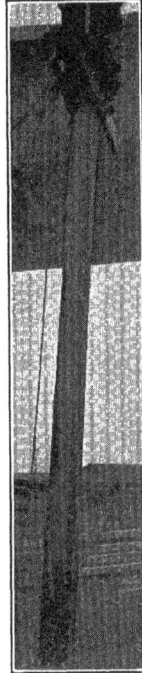

and strength of the construction. However, in the case of poles, where the concrete is deposited within narrow forms, other conditions partly modify or control the distribution. If the metal is concentrated in four equal areas, a rod at each corner, a square pole will be equally strong either parallel or normal to the line.

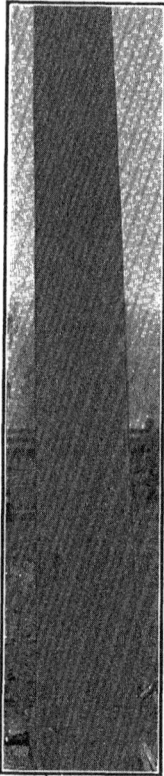

Other or finer distribution of metal with equal strength in both directions necessitates an excess of material over that required for the forces normal to the line. When the metal is concentrated, the fabrication of the reinforcement into a unit frame and also the concreting operations, are more easily accomplished. It may be said, as in the case of beams, that ample web reinforcement assures a firm unyielding unit during concreting, as well as provision against vertical shearing stresses.

In other fields of reinforced concrete work high-carbon steel with a high elastic limit and a correspondingly richer concrete are being used to permit high working stresses in design. If, in such work, high-unit stresses can be used, with a percentage for impact, it seems entirely reasonable to allow correspondingly high working stresses in pole design, since the severe conditions of loading occur infrequently.

In the construction of concrete poles or other structures in which there is a relatively large and important amount of reinforcing, great care must be exercised to thoroughly tamp or puddle the concrete as it is deposited, in order to prevent pockets and to insure every lineal inch of metal having a firm adherence to the concrete. In such structures, the increase in stress in the reinforcement must be very rapid and such additions of stress are dependent on the efficiency of the connection between the steel and the

FIG. 95.—Well-designed concrete pole tested to destruction, showing multiplicity of cracks.

concrete. Mechanical bond or deformed bars, *i.e.*, twisted squares or bars with various projections on their surfaces, are superior to smooth bars for work in which high stresses must be developed in short lengths. Rods may often be bent into hooks or clamped together with advantage.

Reinforcing metal may be either medium-grade steel having an ultimate strength of 60,000 to 70,000 lb. per square inch, and an elastic limit of 30,000 to 40,000 lb. per square inch and capable of being bent cold about its own diameter, or it may be high-carbon steel of 80,000 to 100,000 lb. per square inch and an elastic limit of 40,000 to 60,000 lb. per square inch and capable

FIG. 96.—Concrete telephone and telegraph poles.

of being bent cold without fracture about a radius equal to four times the diameter of the rod. Since the elastic limits of these two grades of material are quite different and since the failure of a concrete pole occurs when the tension surface cracks, there will be no similarity between two poles of the same dimensions and amount of reinforcement in which different grade rods are

11

used. Owing to the fact that in a pole the stresses in the reinforcement must change rapidly in amount with every lineal foot
of the pole, it is most essential, at least for high-strength poles,
to use mechanical bond or twisted bars. It is also necessary to

FIG. 97.—Loading concrete poles.

FIG. 98.—Hollow concrete poles.

provide diagonal or spiral reinforcing when poles are to be subjected to torsion, although close spacing of horizontal ties will
be of assistance. Horizontal ties are needed primarily to restrain the rods from local buckling with its consequent spalling

off of concrete. The rods must be tied to the horizontal straps or other secondary system at each intersection, in order to assist in developing bond stresses. In view of the character of service

FIG. 99.—Concrete distribution line poles.

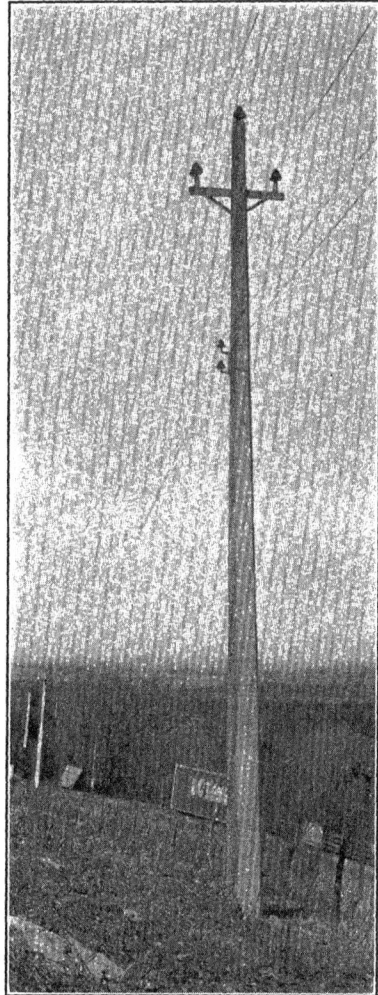

FIG. 100.—Concrete transmission line poles.

to which horizontal bands or spacers are subjected, it is utterly indefensible to use cast rings or bands.

No attempt should be made to remove the forms until the concrete has obtained a good set and care must be exercised even then to prevent injuring the surfaces during such removal. The forms should be kept covered during setting, particularly when exposed to direct sunlight in hot weather. After the forms have been removed the concrete pole should be well sprinkled and kept under canvas for some days. A freshly made pole cannot be handled or rolled with impunity until it has become well set. Further, the subsequent handling, particularly of long poles, must be done with care and preferably with slings attached at two separate points.

The bolt holes and step-bolt sockets must be cast in place. Hardwood blocks may be used for step bolts, but a cast or spiral socket is preferable. Plastering the surface of poles to remove pockets or to produce a finished surface is particularly objectionable. The former should be avoided by proper workmanship in the first place. The latter is entirely unnecessary since a very fine surface can be readily produced by rubbing.

The most commonly used mixture is 1:2:4 Portland cement, sand, and broken stone or gravel. It should be mixed wet, using carefully selected materials, and tamped or churned to eliminate air bubbles, obtain a good surface, and thorough contact with the reinforcement. Such a mixture when well made has an average compressive strength of about 1000 lb. per square inch in seven days, 2400 lb. per square inch in one month, 3000 lb. per square inch in three months, and 3500 lb. per square inch in six months. If conditions make it desirable to use high working stresses, a month or more should be allowed to elapse before new poles are subjected to severe tests.

CHAPTER IX

FOUNDATIONS

The proper penetration of wood poles is the result of many years of actual experience under the varying conditions of different soils. It is not a matter which can be determined accurately and readily by a mathematical formula. In an ideal

FIG. 101.

formula there must be a variable "constant," the value of which depends on the particular soil in which the poles are to be erected. Inasmuch as it is impracticable to make preliminary tests of the soil conditions on a long transmission line, it is necessary either to have several standards or to design a single foundation which will provide safely for ordinary variations in the soil.

In an attempt to reduce the initial cost of installations a number of lines have been built with weak foundations. Some of these supports have already failed and the length of service of others is a matter of conjecture. It is true that the cost of foundations for wide-base structures may be comparatively high, but the insurance value of a good foundation justifies its cost. The use of shallow foundations is doubtless due in a measure to the methods in vogue in testing towers. A test tower on a concrete or metal foundation not only gives no information in regard to the subsequent foundation, but induces a false sense of security both in the probable action of the foundation and of the tower itself.

Fig. 102.—Bog shoe.

A foundation has two functions to perform: first, to prevent uplift or depression, and second, to resist buckling at the point of maximum leverage. Therefore, a foundation member or ground stub which is not firmly braced against horizontal movement at the ground line may introduce stresses not contemplated by the designer. A test tower on a rigid foundation will give higher test values than the line towers in actual service, and the test is, therefore, deceptive by the amount representing the effect of this rigidity. In structures in which the main legs are unsupported for relatively large values of l/r, or in which the tower itself has much less strength in one direction than in the other, a rigid anchorage is necessary to provide the conditions assumed in the design. In some soils certain types of poles and towers have a strength in excess of that of the foundation, the result being that the weakest part of the line is not even suspected until failure occurs. There are no pole penetrations of less than 5 ft.

in the standard pole settings for wooden poles, but some metal structures have been designed with penetrations of 3 ft. 6 in., which is not as deep as the ordinary frost line.

On the other hand, the enlarged butts of latticed steel poles encased in concrete would warrant some reduction in their depth of penetration as compared with that of similar wood poles not so incased.

The protection of the metal in the anchorage is naturally of vital importance to the permanence of the structure. If there are some localities in which galvanizing is no real protection, there are other places where galvanizing is more economical

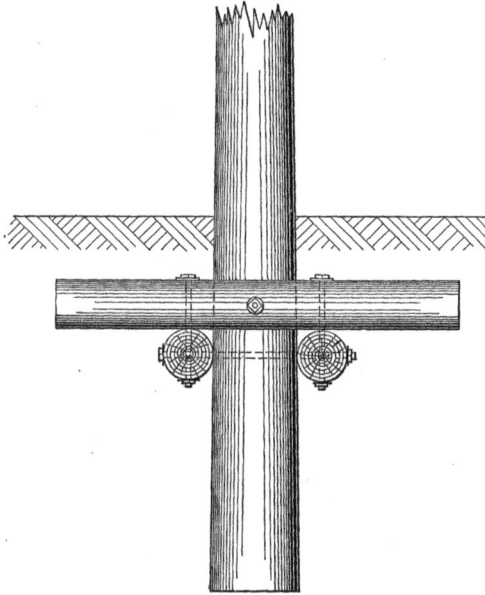

Fig. 103.—Bog shoe.

than paint. When the anchorage is incased in concrete the incased portion may safely be considered as having a longer life than the superstructure. The point of entrance of the metal in the concrete is usually considered as the location of the future maximum corrosion, whether the structure is painted or galvanized. The writer believes that this assumption is not always correct, and that the location of future deterioration will depend on the relative effect of acids, etc., carried by air currents to the upper portions of the structure, *versus* the amount of dirt and water at the base. However, there can be no question of the propriety of protecting ground metal, and data are needed on the results obtained by galvanizing, concrete, asphalt, tar, treated burlap, additional coats of paint, etc. With poles and towers as in any other kind of construction, a poor foundation induces a poor superstructure, and it is probable that more tower failures have been caused or superinduced by faulty foundation design than by any other cause.

Wood poles should be set in the ground to depths not less than those specified in the following table:

TABLE 29.—WOOD-POLE SETTINGS

Length over all (feet)	Straight lines (feet)	Curves, corners, dead ends, etc. (feet)
30	5.0	6.0
35	5.5	6.0
40	6.0	6.5
45	6.5	7.0
50	6.5	7.0
55	7.0	7.5
60	7.0	7.5
65	7.5	8.0
70	7.5	8.0
75	8.0	8.5
80	8.0	8.5

In rock excavation smaller penetrations may be permitted, particularly if the back-fill is of concrete. If the filling is of earth, or the fine waste from the excavation, it should be well tamped into place. The character of the rock must be taken into consideration in decreasing the standard settings, since rotten rock with a horizontal or inclined cleavage may be but little stronger than the better grades of packed earth.

Although the vast majority of pole foundations are included in the very elastic term "earth," it is occasionally necessary to set poles in bad ground. There are many varieties of uncertain soils and no one design is absolutely adequate for all conditions.

Light shifting sand, peat bog, wet clay and black muck are of different genera and frequently of different species. The problem usually includes the necessity of providing for a lack of proper lateral resistance at the ground line, and it may or may not include a protection against settlement.

The simplest foundation reinforcement, and one also used to strengthen a heavily stressed pole in good ground is the provision of a crib or toe and heel braces. A common error in constructing such braces is that of a weak connection to the pole. Inasmuch as the ground on one side of the pole is not exactly like that against which one end of the brace presses, there is always some tendency to rotate. If a high resistance must be developed, it generally requires very considerable earth pressures over a

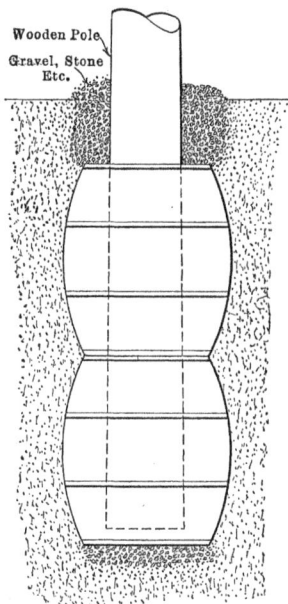

FIG. 104.—Barrel foundation.   FIG. 105.

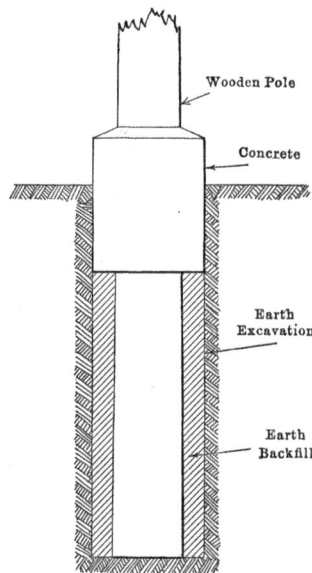

restricted area. The earth pressures are available if properly developed, and they are much greater than is often supposed, the writer having seen 1-in. bolts sheared through the wood across the grain of a 3-in. yellow-pine plank, by the pressure from a bank of wet gravelly clay.

Anti-tilting platforms or crib work are sometimes entirely satisfactory when placed 4 or 5 ft. below ground, or at about one-half the penetration of the pole. These platforms serve a triple purpose in that they prevent tilting, settlement, and uplift. To be effective against tilting, they must, of course, be securely fastened to the pole, either at two elevations or partly by means of inclined hanger rods.

Concrete heel and toe braces have been advocated, but it seems inconsistent to use concrete as a brace for timber, unless it extends above the ground line as a protection against decay. If, however, an envelope of concrete extending 1 to 2 ft. above and below the ground line is to be used, it would be advisable and economical to form the braces of concrete. In the writer's opinion a short envelope of concrete like that just mentioned is very desirable for important wood pole lines where the cost of handling water and aggregates is not prohib-

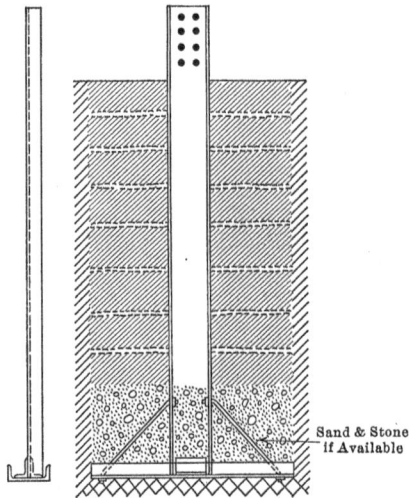

Fig. 106.—A-frame foundation (semi-obsolete type).

itive. This design has been used very little, but it appears to possess considerable merit, as it protects the timber from attack by grass fires, and decay at the most exposed point, and further-more increases the lateral stability of the pole by increasing its diameter where the earth resistance is weakest.

Poles set in stout well-hooped headless barrels filled with sand and gravel have been used with success in peat bogs, while a removable barrel form has also been used to deposit sand and gravel as back filling. The permanent barrel forms a large butt and retains the filling which might otherwise work away from the pole through the surrounding earth. In case the subsoil is compact, the barrel may not be required, sufficient resistance being supplied by the back filling.

It has always seemed to the writer that shallow foundations were particularly objectionable for flexible frames and that instead of a penetration of 5 ft., not less than 6 to 7 ft. should be used. When the ground is loosened in the spring or fall, either by thawing or rains, the upper 1 or 2 ft. are of relatively little service in resisting uplift or lateral pressure.

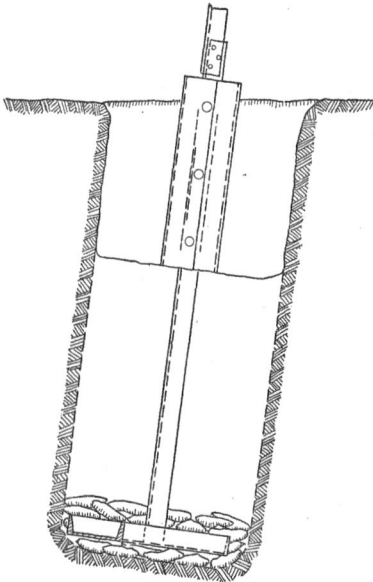

Fig. 107.—Tower foundation, concrete-filled sleeve.

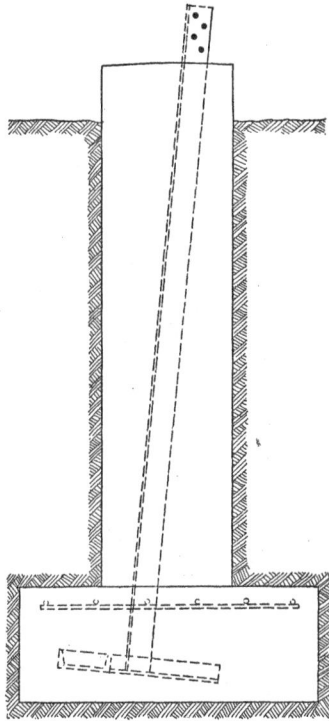

Fig. 108.—Wide-base tower foundation.

Unless the bottom of an earth excavation is naturally very firm, about 6 in. of mixed sand and gravel, or sand and broken stone should be tamped into the bottom of the hole to form a firm bearing for the anchor plates. The earth back-fill above the anchor plates should be tamped in thin layers. Concrete back filling is, of course, preferable, but its cost may be considerable when it is necessary to haul water and cement and perhaps the sand and gravel.

It must be admitted that exact and concise rules for foundation design are not available, and this condition has lead to inadequate and also to very extravagant designs. The latter may in part be due to the assumption of unnecessarily severe condi-

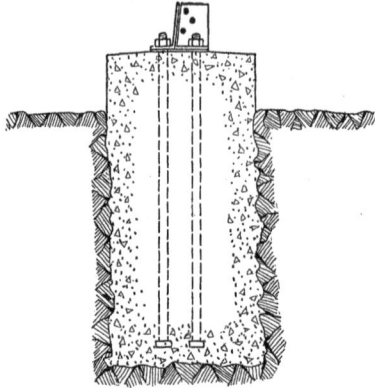

Fig. 109.—Tower foundation in rotten rock.

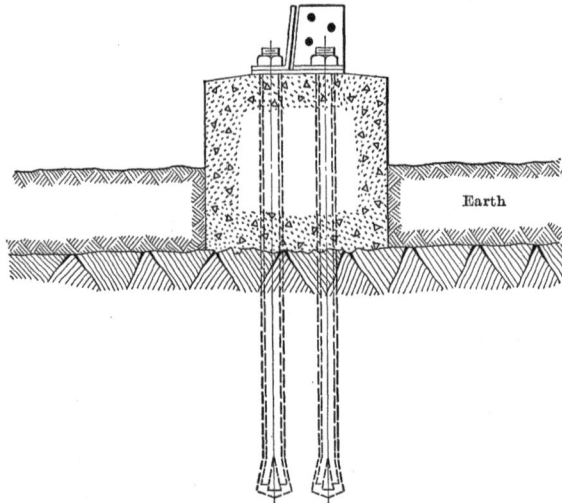

Fig. 110.—Tower foundation in rock.

tions of loading. Thus it will be found that the foundation for a small-base steel tower or pole may become excessive if designed for a factor of safety of 2.0 for all wires broken and using as a resisting force only the weight of the 30° cone of earth.

Transmission-line experience tends to show that the actual resistance of average ground must be greater than that given by the usual formulas. This may be due to the probability that the specified maximum load occurs when the ground is frozen or partly frozen and that under such conditions high resisting stresses are developed.

The assumption that the resistance is merely that of the weight of the earth inclosed within the inverted cone outlined by the angle of repose of the material, assumed as 30° or even 45°, cannot be correct for compact soil, clay, etc. It would seem that good ground, particularly if bound together by roots, stones, etc., must have an appreciable shearing value.

CONCRETE

Cement.—The production of Portland cement in this country is now so standardized that any brand of Portland cement will pass the standard tests, provided the cement in question is a representative sample of the maker's output. The chief purpose of cement specifications is, therefore, to secure the proper quality rather than to discriminate between brands. It should not be understood that all cement is necessarily satisfactory, since any particular lot may have been injured by an error in manufacture or by improper or overlong storage. Any reputable manufacturer can and does make a satisfactory cement and the cement tests should, therefore, be used merely to guard against error. The manipulation of the tests and the requirements to be met by the cement have been very completely standardized and are given in what are known as the "Standard Specifications for Cement," copies of which may be obtained from any cement manufacturer or from the Association of Portland Cement Manufacturers, so they will not be repeated here.

Proportions.—The proportions of cement, sand and stone, or gravel, will depend somewhat on the purpose for which the concrete is to be used. In general, the smaller the volume and the greater the stresses, the richer the concrete should be. Further, if the concrete is to be impervious to water or is to be immersed in water, or deposited through water, the mixture should be richer, *i.e.*, with a larger amount of cement than would otherwise be necessary. Concrete is stronger, more impervious, and permanent when it is of the maximum density. The maximum density obtainable from any given sand, cement and stone, or gravel, will be that due to one certain proportion of the ingredients. The proper proportions in any particular case will be determinable by tests designed to disclose the voids in the aggregates. Ordinarily, however, it is not necessary to make such tests, as the customary proportions, combined with good workmanship, will produce a satisfactory result. The amounts of sand and stone have usually been given separately, although in reality there should be two proportions—that of the cement and that of the combined sand and stone. The most commonly used proportions are: 1:6 (1:2:4) for fine work, and 1:9 (1:3:6) for mass foundations, etc.

Aggregates.—The aggregates, which are the sand, gravel, broken stone, slag, cinders, chats, etc., may be of various sizes from screenings to fairly large stones. They should, however, be of graded sizes in order to present fewer voids. Inasmuch as concrete is in reality an artificial stone, its constituent parts must be free from vegetable matter and soft particles, or the resulting product will be in the nature of "rotten rock." It is frequently specified that the sand and other aggregates shall be clean, although a small percentage of clay is generally permissible, since neither sand nor gravel will be perfectly clean without very thorough washing. It has also been required that the broken stone shall be sharp, but this is generally not necessary since a high grade of concrete can be made with gravel, and gravel is never sharp. Sharp sand, however, is desirable.

Water.—The water used in mixing concrete should be free from oils, acids, or any considerable amounts of alkali or vegetable matter. Satisfactory water can generally be obtained throughout the country, and usually near the site of the work. Although "dry concrete" has been used to a considerable extent abroad, and was formerly used somewhat in this country, the present practice here is to use "wet concrete." By wet concrete is meant concrete mixed with sufficient water to be semi-fluid, so that it may readily flow around the reinforcing or incased material, and be easily tamped or puddled. This is necessary to completely fill the forms, and obtain an efficient adherence to the reinforcement. The only objection to the use of an excess of water is that some of the cement will be washed away or deposited separately, and that the resulting concrete sets and dries more slowly, thus delaying the work. Since water is needed both for fluidity and chemical combination a sufficient quantity must be provided to prevent its absorption by, or drying on, the aggregates. In warm weather particularly, it is advisable to thoroughly "wet down" the pile of stone from which the material is taken.

Mixing and Placing.—Concrete may be mixed either by hand or by mechanical mixers, the method in any instance depending on the quantity to be made and the availability of a mixer at the site. Machine mixing is probably more thorough than hand mixing, although just as good concrete can be made by hand under proper supervision. In hand mixing, the materials should be mixed on a flat form or floor to prevent an excessive loss of cement-bearing water, or the admixture of earth, etc. Mixing floors are of various sizes from about 6 ft. square up to much larger areas, but in any case it is desirable that tongue-and-grooved lumber, or two layers of lumber, be used to decrease the leakage through the cracks.

In order to obtain the proper proportions of the aggregates some unit of measurement such as a bucket or barrow should be used to transport them from the stock piles to the mixing platform. It is then a simple matter for the workmen to regularly take so many units of sand, another number of units of stone and the specified number of bags of cement to make each batch of concrete. The sand and stone are first placed on the mixing board and mixed by turning the pile over with shovels. The cement is then spread over the mass which is turned, water being added during the turning. The number of turns to be given each process, and the faithfulness with which the work is done determines the excellence of the mixing. Mixing should be continued until the mass presents a uniform appearance, and the stone appears to be entirely covered with sand and cement. As soon as the mixing is completed, the material should be taken in water-tight buckets or barrows, and placed in the work. The size of the batch should depend on the amount that can be immediately used, since material left on the board for any considerable time takes an initial set, and is useless for future work.

The placing of concrete should be as nearly continuous as practicable to prevent the formation of cleavage planes or planes having little cohesion. Such surfaces will contain a layer of "dead" material as well as a certain amount of dirt which floats to the surface. In poor grades of work the joints can be distinctly seen on the sides of the structure. Since it is not always possible to work continuously, the temporary surface should be left rough and should be thoroughly washed and preferably scrubbed before continuing operations. Reinforced-concrete poles should always be made in one operation as they are too small to justify the risks attending non-continuous work.

Forms.—The finish and general excellence of the forms should depend on the character and magnitude of the work. In case a ground stub or the base of a wood or steel pole is to be incased in concrete, no forms will be necessary as the earth will serve the same purpose. A small collar may be used, however, at the ground level to retain the concrete above the ground. Forms are generally of wood, but in case there is considerable repetition in the construction, it will often prove economical to use metal, as metal forms last longer, retain their shape, are more readily cleaned and produce a smoother finish on the surface of the concrete.

The thickness of the lumber and the amount of bracing necessary for wooden forms will depend on the size and shape of the structure. Where the same form is to be used again and again, it will prove economical to use thicker material and better made forms than for a single unit, since thin or flimsy forms become warped and broken in handling. The inner surface of the form should be of faced lumber and fillets should be placed in the corners, since sharp edges are difficult to make and rarely permanent. On pole foundations the corners above ground should be beveled or worked into a curve.

The proper time for the removal of forms will vary considerably, depending on the temperature and the kind of cement. Concrete sets slowly in cold weather and also under certain conditions of the atmosphere may take a much longer period than is usually required. This delayed set is not due to freezing or very cold weather, and its exact nature is not well understood. There seems to be nothing to do but to wait until some slight change in the temperature and humidity will allow the concrete to take its usual set.

In order to prevent the adherence of concrete to the forms with the consequent danger of breaking out pockets on the surface when the forms are removed, they may be coated on the interior with soap, grease, etc. After the forms are removed and in some cases even before removal, the concrete should be kept damp by sprinkling. This is particularly desirable in warm weather to prevent the formation of surface cracks.

Workmanship.—As indicated above the character of the workmanship may properly vary for different kinds of construction, that for simple and massive foundations being of a lower grade than would be necessary for reinforced concrete in small volume particularly such work as reinforced-concrete poles. Poles require the very best possible grade of workmanship and material, as well as competent designing and supervision.

It is not always possible to avoid concrete construction in cold weather, but if the temperature is much below freezing or even slightly below for continued periods, it is necessary to adopt some method to prevent the concrete from freezing before it has set. While frozen concrete will usually set after it has thawed, there is a possibility of its failing if loads are applied while it is still frozen. When the temperature during the day is above freezing, even if the nights are cold, the heat generated by the concrete, will often be sufficient to prevent freezing. Otherwise, it becomes necessary to heat the sand and stone before mixing and to enclose the concrete with planks, canvas, straw, etc., or to heat the enclosure by fires or steam.

Forms serving as temporary supports for the concrete should not be removed too promptly. Pole and tower foundations, however, may have their forms removed rather quickly, since the forms do not usually serve to support the concrete. The early removal of forms is often due to the desire to rub the exposed surfaces before the concrete has become very hard. It is entirely possible, however, with the use of, perhaps, a little additional energy, to rub to a smooth finish surfaces which have set for several weeks.

It is generally specified that surfaces shall not be finished by plastering on a coating of cement mortar, but this requirement is not always properly enforced. Since such plastered material will frequently spall off, it is far better to rub the surface with a concrete brick or carborundum stone, using water instead of mortar. By this means, the marks of the forms as well as an outer layer of cement are removed leaving a permanent sandstonelike finish.

Reinforcement.—Although some specifications, notably those of the Joint Committee on Concrete and Reinforced Concrete, have permitted only medium-grade steel, there appears to be

12

absolutely no necessity of adhering to this requirement in transmission-line construction provided high carbon steels are used with proper design and inspection. Open-hearth high-carbon steel rods are obtainable, as well as those of rerolled rail stock, with an ultimate strength of 80,000 to 100,000 lb. per square inch, and an elastic limit of 40,000 to 60,000 lb. per square inch, and which are capable of being bent cold without fracture about a radius equal to four times the diameter of the rod.

Such material cannot be subjected to the same amount of abuse in handling as the low-carbon steels, and care is necessary in testing and inspection to insure that any given lot of rods will have about the same characteristics. When intelligently used, however, such material is entirely satisfactory for certain classes of transmission-line construction.

The system of reinforcement should provide material designed to prevent or at least minimize cracks due to shrinkage. In general this is accomplished by the use of both longitudinal and transverse reinforcing metal. While in some instances shrinkage cracks may not impair the strength of the construction they might, by the admission of water, cause corrosion of the reinforcing metal.

The reinforcement should not be composed of a few large bars when it is possible to use smaller rods without a material reduction in the effective lever arm. Very small rods are not always desirable, but the use of two medium-sized rods in place of one very large one allows greater uniformity, and a maximum adherence of the concrete to the skeleton.

Waterproofing, Salt Water, Alkali, Etc.—Waterproofing of transmission-line structures will usually be confined to the exclusion of water from the incased reinforcing rods or structural steel. It seems probable that the most convenient form of protection will be a dense mixture and a firm uncracked surface. In other classes of construction, waterproofing by the various tar and felt processes is efficient when well done, and in some cases the addition of waterproofing compounds to the cement has given satisfactory service. In general, however, the same amount of care if expended in making proper concrete will presumably provide equal protection to line structures. The thickness of concrete outside any rod or embedded metal must be sufficient to minimize the admission of water through hair cracks or by capillary action. In concrete poles the outer thickness of concrete will rarely exceed $1\frac{1}{2}$ in. whereas in other construction 6 in. can be given at little additional expense.

Concrete which is to be immersed in sea water must be made of richer mixtures than otherwise required and care is essential to secure a dense mixture and an impervious surface. If the concrete can be allowed to become hard before the sea water is permitted to come into contact with it, its resistance to disintegration will be increased. In many instances the disintegration of concrete exposed to sea water has unquestionably been due in part to mechanical abrasion from floating ice, logs, etc. Added to this there has been a minute spalling of the surface due to freezing in the surface. The chemical attack by sea water is believed to be caused by the replacement of a part of the CaO of the cement by MgO from the sea water, and also by a change in the proportions of silica and SO_3. The most effective protection against sea water is believed to be provided by a properly proportioned very dense concrete, rather than by the addition of any of the various materials which are intended to reduce permeability.

In some sections of the country, particularly in certain western states, the soil or ground water may contain sufficient acid or alkali to affect concrete unless the concrete is carefully made. As in the case of waterproofing or salt water concrete, the most effective protection against acid or alkali is dense concrete, although this should be supplemented by the provision of aggregates inert to the acids or alkalis. If cinders are used as an aggregate, they should not contain much sulphur, and should be hard and fairly non-porous.

Accurate and complete data are lacking in relation to the service of concrete subjected to electric currents, but it is probable that an embedded rod which acts as the anode of an electric circuit is liable to corrode and to disrupt the concrete in contact with it. Since corrosion caused by electrolysis is confined to the surface from which a current flows, it would appear practicable to remove any liability to disintegration in line structures by grounding them and, perhaps, by allowing the incased material to project through the bottom of the concrete into the earth.

CHAPTER X

PROTECTIVE COATINGS

Paint and Painting.—The careless application of paint and the use of inferior grades of paint on transmission line structures cannot be too strongly condemned. Although paints range in price from about 40 cts. to $1.40 per gallon, the possible saving afforded by using the lower priced paints is small compared with the risk of deterioration. The usual requirements for painting are few in number and impossible of misinterpretation, but unfortunately they are usually more or less disregarded.

Inasmuch as the adherence of the first coat of paint, as well as all subsequent coats, depends on the manner in which the first coat is applied and on the character of the surface to which it is applied, inspection should begin with the shop coat. Careful inspection of such work is, however, the exception rather than the rule. All painting specifications require that the surfaces shall be *thoroughly* cleaned of all mill scale, rust and dirt, and that no painting shall be done in the rain. It is to be inferred that exposing fresh paint to rain and cinders from a railroad or shipping yard would be equally undesirable. Unfortunately, most manufacturers have but meager covered-storage facilities and a protracted period of rain, or insistence on the part of purchasers that shipments be made immediately, will often result in undue exposure of freshly painted structures. Further, the application of the second or field coat is frequently open to improvement, in that careless workmen will paint over mud spots or blisters. Since the top coat has no adherence to the metal at such points, it will flake off when the mud loosens or the blister cracks. If the shop coat is in bad condition in any place it should be scraped clean and repainted. Some muscular effort is necessary to a proper application of paint and the cost of worn-out brushes will be amply repaid by the ultimate result.

The use of thinners in general or the addition of benzine to ready-mixed paints may effect a little saving in the initial cost of labor and material, but will not be economical in final cost.

180

The field coat should be applied after the wires have been strung so that the final coat will not be marred by erection operations. Sometimes one shop coat and two field coats are applied, but it seems that one of these coats might preferably be postponed for a few years, and a portion of its cost utilized in improving the initial painting.

It is the writer's belief that for structures with small or medium-spread bases reasonably thick painted material will usually prove more satisfactory and economical than thin galvanized sections, provided subsequent maintenance is not grossly neglected. As transmission line supports are widely scattered, it has not been customary for the linemen or patrol men to pay much attention to the physical condition of metal structures, with the result that maintenance painting has been postponed until quite noticeable rusting has taken place.

The use of a shop coat of linseed oil instead of paint, although not uncommon some years ago in bridge work, is rarely found in transmission line construction, nor does there appear any good reason for its use. Such coats unless carefully applied are very liable to blister and peel either before or after the field coat is applied. Further, such primers add nothing to the adherence of the final coats, and are of far less service than even a mediocre shop coat.

In repainting, *i.e.*, maintenance painting, just as much, indeed more, care is needed than in the original applications. This is due to the fact that in addition to removing dirt, etc., it is necessary to remove paint if rusting has occurred beneath it, since the new covering of paint will not entirely prevent the continued action of the rust. As in the original painting, no efficient service need be expected of paint applied to a wet surface.

All paints should be stirred before using, particularly the heavy pigment paints such as red lead. A good rule is to require the painters to stir the barrel thoroughly before replenishing any individual can. Barrels which have been opened should be kept, not only under cover, but covered, otherwise water, dirt and other substances become admixed with the paint, or else the paint becomes dried out, thick and dead. The writer is inclined to heartily recommend a plaster of "paint skins" about the point of entrance of the superstructure into the concrete, but no one would advocate an attempt to spread such a thick mortar-like substance over the large areas of the superstructure. Even under rather careful supervision, the bottom of the barrel is a thicker more heavily pigmented paint than the half first used. If to the thick sediment remaining in the lowest quarter barrel—due perhaps in part to evaporation—benzine be added, the result cannot in justice to the manufacturer be termed an equivalent paint. In general, the brand of paint or type of pigment is of far less importance in the ultimate result than a little care and intelligence in its application. To go further, even the "intelligence" might be omitted since ignorance should be able to rub dirt and rust off and rub paint on.

CONDITION OF PAINTED STRUCTURES AS REPORTED IN 1915 BY SIX LINES
HAVING A TOTAL OF 1686 TOWERS IN SERVICE

Number of structures	Year built	Foundation	Condition as reported in 1915
697	1909	Concrete	Fine condition, have not been repainted.
127	1911	Galvanized	Very good, slight rust under paint in spots, due to careless cleaning.
78	1912	Concrete	Slight corrosion.
364	1912	Concrete	Corrosion, only a shop coat was put on and rusting occurred before first field coat was added in 1914.
262	1913	. . . .	Very little corrosion.
158	1913	Galvanized	No corrosion.
1686			

Galvanizing.—Under favorable climatic conditions, particularly over rough unsettled country where all maintenance operations are expensive, galvanizing is thought to be more economical than painting, at least for wide base towers. In the more compact structures the cost of painting is reduced owing to the ease with which a painter can move about. No very definite knowledge exists as to the regions in which galvanizing does not afford proper protection, although it is generally assumed as being unsatisfactory in the neighborhood of coke ovens, smelters, steam plants, etc., and near the seacoast. It may also be true that certain soils such as swamps, induce rapid deterioration. It should be noted that the use of sections $\frac{1}{8}$ in. in thickness presupposes that their protection by galvanizing will be absolutely effective. The galvanizing of structural members—by the hot process—can be done in a very uniform manner and in strict accordance with the standard requirement, even though the real efficiency of the average inspection is open to question. The standard test is an accelerated specimen test, and although the only practicable one, is subject to the usual criticisms of such tests.

CONDITION OF GALVANIZED STRUCTURES AS REPORTED IN 1915 BY NINETEEN LINES HAVING A TOTAL OF 9269 TOWERS IN SERVICE

Number of structures	Year built	Foundation	Condition as reported in 1915
184	1906	Earth	Good condition.
160	1906	Earth	Fair condition.
111	1909	Earth	No corrosion.
244	1909	Earth	Good condition.
913	1909	Painted	Some corrosion when exposed to salt spray and salt fogs.
220	1911	Earth	No particular signs of rust, except on sherardized bolts.
242	1911	Earth	No corrosion.
82	1911	Concrete	Entirely satisfactory, last inspection.
378	1911	Concrete	Good condition.
748	1911	Concrete	Good condition.
593	1912	Earth	No signs of corrosion.
1041	1912	Concrete	No signs of corrosion.
324	1912	Painted	Unimpaired.
1852	1912–1914	Earth	Good condition.
1079	1913	Earth	Good condition.
33	1913	Concrete	No corrosion.
851	1913	Concrete	No corrosion.
64	1913	Concrete	No corrosion.
150	1913	Earth	No particular signs of rust, except sherardized bolts.
9269			

The ideal galvanizing consists in a thick, tough layer of zinc absolutely free from pin holes and adhering firmly to the steel. A pure high-grade coating of zinc is more likely to be tough and to have a good adherence to the steel than one which is composed of lower grade material. The excellence of the pickling and cleaning will have a marked effect upon the existence of pin holes, since these are generally caused by more or less minute specks of scale, etc. The efficiency of the final work may, therefore, depend either on the material of the bath or on the workmanship.

In order to obtain adherence, the steel is first pickled in a weak solution of sulphuric acid to expose a clean surface. The pickling may or may not be followed by washing, but it should be followed by an examination and removal of spots. The next step is the muriatic acid bath after which the metal is heated either in an oven or by placing it over the bath of hot zinc.

The material to be coated is then immersed in the hot bath of molten zinc, and allowed to remain therein until in the judgment of the foreman the covering is complete. Before entirely removing the material from the bath it is allowed to drain to recover the excess zinc. This process is the cause of the small projections, or icicles, of zinc which are frequently to be found on galvanized structural material. Apart from the question of appearance, they do no harm, unless at contact surfaces, and are to some extent at least the sign of a thick bath.

The entrance end of the zinc bath contains a floating bath of sal ammoniac retained in position by movable plates set vertically in the bath. All entering material is inserted through this "flux."

The exit end of the zinc bath or "kettle" may contain an admixture of "temper metal," added for the purpose of increasing the luster and insuring a more fluid bath. This added material contains various proportions of tin, aluminum and spelter. The addition of tin produces the spangled appearance used for galvanized buckets, boilers, etc., and this ingredient will generally be found in the temper metal of baths used for general commercial galvanizing. The amount of tin, however, should be small, and if used it should be placed only in the exit end of the bath, since an excessive quantity, particularly in the inside layer, may result in the formation of a more brittle coating.

If there is any mill scale on the black material, which is not loosened and removed in the pickle or before galvanizing, it will result in a blotch of spelter which may or may not become loosened subsequently.

Owing to the restricted size of the hot bath, although some of them are about 3 ft. square and 25 ft. long, it is necessary to dip long members from each end. This often results in a lap or wave in the coating at the middle of the piece, which is not in itself objectionable. The chief criticism of double end dipping is that it often causes warping of certain kinds of material such as wide plates or thin flexible pieces.

The spelter to be used in galvanizing wire should be that known as High Grade, and the spelter for galvanizing structural shapes should be Prime Western, or equal.

It has frequently been claimed that paint will not adhere to a galvanized coat. Whatever the fact with respect to subsequent painting with structural paints, there is no difficulty in painting the assembling and shipping marks on the finished members with some interior paints.

Galvanizing is one of the few operations in which there has been no considerable attempt either to develop complete methods of inspection, or to make any inspection of the spelter material as such. In view of the different qualities of spelter obtainable it would seem advisable to periodically take samples from the middle of the bath and subject them to a chemical analysis. The following material specification represents fair commercial grades of spelter suitable for wire and structural galvanizing.

ABSTRACT FROM STANDARD SPECIFICATIONS FOR SPELTER

American Society for Testing Materials. Adopted August 21, 1911

Under these specifications Virgin Spelter, that is, spelter made from ore or similar raw material by a process of reduction and distillation and not produced from reworked metal, is considered.

A............................ HIGH GRADE.
D............................ PRIME WESTERN.

A brand shall be cast in each slab by which the maker and grade can be identified.

The maker shall use care to have each carload of as uniform quality as possible.

A. HIGH GRADE.—The spelter shall not contain over

0.07 per cent. lead.
0.03 per cent. iron.
0.05 per cent. cadmium.

It shall be free from aluminum.

The sum of the lead, iron and cadmium shall not exceed 0.10 per cent.

D. PRIME WESTERN.—The spelter shall not contain over

1.50 per cent. lead.
0.08 per cent. iron.

The slabs shall be reasonably free from surface corrosion or adhering foreign matter.

No less than ten slabs shall be taken as a sample from each car; for smaller lots, in the same proportion to the total number, but in no case less than three slabs. In case of dispute half of the sample is to be taken by the maker and half by the purchaser; and the whole shall be mixed.

The slabs selected as samples are to be sawed completely across and the sawdust used as a sample. In case no saw is available for this purpose, the slabs should be drilled completely through and the drillings cut up into short lengths. The saw or drill used must be thoroughly cleaned. No lubricant shall be used in either sawing or drilling and the sawdust or drilling must be carefully treated with a magnet to remove any particles of iron derived from the tools.

LEAD.—For the determination of lead in High Grade not less than 25 grams, and in Prime Western not less than 5 grams shall be taken; that is, the sample used for analysis should not contain less than 0.01 gram lead.

IRON.—The sample for iron should contain not less than 25 grams for the three higher grades and not less than 10 grams for Prime Western. The entire sample must be dissolved, the iron precipitated as ferric-hydroxide, then redissolved, reduced, and the iron determined by titration.

CADMIUM.—Dissolve 25 grams in 330 cc. of a solution of one part of hydrochloric acid (specific gravity 1.2) and five parts of water. Let it stand over night; filter and wash; reject filtrate and dissolve the residue, which should be about 5 per cent. of the zinc, in nitric acid. Add 10 cc. of sulphuric acid; evaporate to fumes; dilute and filter out and wash the lead sulphate. Dilute the solution to 500 cc.; add 5 grams of ammonium chloride; pass a slow stream of hydrogen sulphide for one hour and let stand for about five hours; filter, wash with hot water, dissolve in 10 cc. of sulphuric acid and 50 cc. of water; filter and wash. Dilute to 400 cc.; precipitate with hydrogen sulphide as before. Weigh as cadmium sulphide or dissolve in hydrochloric acid and titrate with potassium ferrocyanide.

The following standard specifications for galvanizing are those now adopted by practically all interested parties and are applicable to all galvanized material.

STANDARD SPECIFICATIONS FOR GALVANIZING

(1911 Edition)

These specifications give in detail the test to be applied to galvanized material. All specimens shall be capable of withstanding these tests,

A. Coating.—The galvanizing shall consist of a continuous coating of pure zinc of uniform thickness and so applied that it adheres firmly to the surface of the iron or steel. The finished product shall be smooth.

B. Cleaning.—The samples shall be cleaned before testing, first with carbona, benzine or turpentine and cotton waste (not with a brush) and then thoroughly rinsed in clean water and wiped dry with clean cotton waste.

The samples shall be clean and dry before each immersion in the solution.

C. Solution.—The standard solution of copper sulphate shall consist of commercial copper sulphate crystals dissolved in cold water, about in the proportion of 36 parts, by weight, of crystals to 100 parts, by weight, of water. The solution shall be neutralized by the addition of an excess of chemically pure cupric oxide (CuO). The presence of an excess of cupric oxide will be shown by the sediment of this reagent at the bottom of the containing vessel.

The neutralized solution shall be filtered before using by passing through filter paper. The filtered solution shall have a specific gravity of 1.186 at 65°F. (reading the scale at the level of the solution) at the beginning of each test. In case the filtered solution is high in specific gravity, clean water shall be added to reduce the specific gravity to 1.186 at 65°F. In case the filtered solution is low in specific gravity, filtered solution of a higher specific gravity shall be added to make the specific gravity 1.186 at 65°F.

As soon as the stronger solution is taken from the vessel containing the unfiltered neutralized stock solution, additional crystals and water must be added to the stock solution. An excess of cupric oxide shall always be kept in the unfiltered stock solution.

D. Quantity of Solution.—Wire samples shall be tested in a glass jar of at least two (2) in. inside diameter. The jar without the wire samples shall be filled with standard solution to a depth of at least four (4) in. Hardware samples shall be tested in a glass or earthenware jar containing at least one-half (½) pt. of standard solution for each hardware sample.

Solution shall not be used for more than one series of four immersions.

E. Samples.—Not more than seven wires shall be simultaneously immersed and not more than one sample of galvanized material other than wire shall be immersed in the specified quantity of solution.

The samples shall not be grouped or twisted together, but shall be well separated so as to permit the action of the solution to be uniform upon all immersed portions of the samples.

F. Test.—Clean and dry samples shall be immersed in the required quantity of standard solution in accordance with the following cycle of immersions.

The temperature of the solution shall be maintained between 62° and 68°F. at all times during the following test.

First. Immerse for one minute, wash and wipe dry.

Second. Immerse for one minute, wash and wipe dry.

Third. Immerse for one minute, wash and wipe dry.

Fourth. Immerse for one minute, wash and wipe dry.

After each immersion the samples shall be immediately washed in clean water having a temperature between 62° and 68°F. and wiped dry with cotton waste.

In the case of No. 14 galvanized iron or steel wire, the time of the fourth immersion shall be reduced to one-half minute.

G. REJECTION.—If after the test described in Section "F" there should be a bright metallic copper deposit on the samples, the lot represented by the sample shall be rejected.

Copper deposits on zinc or within 1 in. of the cut end shall not be considered causes for rejection.

In the case of a failure of only one wire in a group of seven wires immersed together, or if there is a reasonable doubt as to the copper deposits, two check tests shall be made on these seven wires and the lot reported in accordance with the majority of the sets of tests.

NOTE.—The equipment necessary for the tests herein outlined is as follows:

Filter paper.

Commercial copper-sulphate crystals.

Chemically pure cupric-oxide (CuO).

Running water.

Warm water or ice as per needs.

Carbona, benzine or turpentine.

Glass jars at least two (2) in. inside diameter by at least four and one-half (4½) in. high.

Glass or earthenware jars for hardware samples.

Vessels for washing samples.

Tray for holding jars of stock solution.

Jars, bottles and porcelain basket for stock solution.

Cotton waste.

Hydrometer cylinder three (3) in. diameter by fifteen (15) in. high.

Thermometer with large Fahrenheit scale correct at 62° and 68°.

Hydrometer correct at 1.186 at 65°F.

CHAPTER XI

LINE MATERIAL

Tie Wires.—The usual function of the wire "pigtails," used to attach the conductors to the insulators, is to prevent falling and creeping rather than to effect dead-ending. Even assuming the most efficient form of tie attachment, there is some unknown relation between the size and number of turns of the tie wire and the size of the conductor which will be dead-ended thereby. Further, the effectiveness of a tie wire is almost entirely dependent on workmanship and the attachment must be made without nicking the power wire. Stranded cable, particularly in large sizes, can be tied more effectively than solid wire, owing to the grip of the tie between the strands.

Tie wires should either be made of the same material as the conductors, or be coated with that material to prevent electrolytic action. The wire itself must be "dead soft" or free from any tendency to spring loose after wrapping. Many kinds of ties have been devised, although they are all variations of two general types, *i.e.*, those depending upon a severe constrictive action in one or two turns, and those in which friction is developed over the extended surface provided by a number of turns. The latter type is generally preferred, especially for soft copper and aluminum, as it is less liable to injure the conductor.

The efficiency of the tie has a very direct bearing on the conditions of loading which can logically be assumed for the poles or towers, although the existence of this relationship seems to have been generally ignored. Thus, it is probable that in very many installations, certain assumptions are made as to the stresses that may be transmitted by the wires to the supports, without any very direct information either as to the type of the tie to be used or its ability to transmit such loads.

For example, if a given tie, or any tie, is incapable of developing a stress of 5000 lb., a No. 0000 copper cable cannot exert its maximum tension on the structure, because the tie would fail before transferring such a load. The example chosen is perhaps

189

unnecessarily severe, since a No. 0000 copper cable cannot be successfully dead-ended by any tie or pin insulator with which

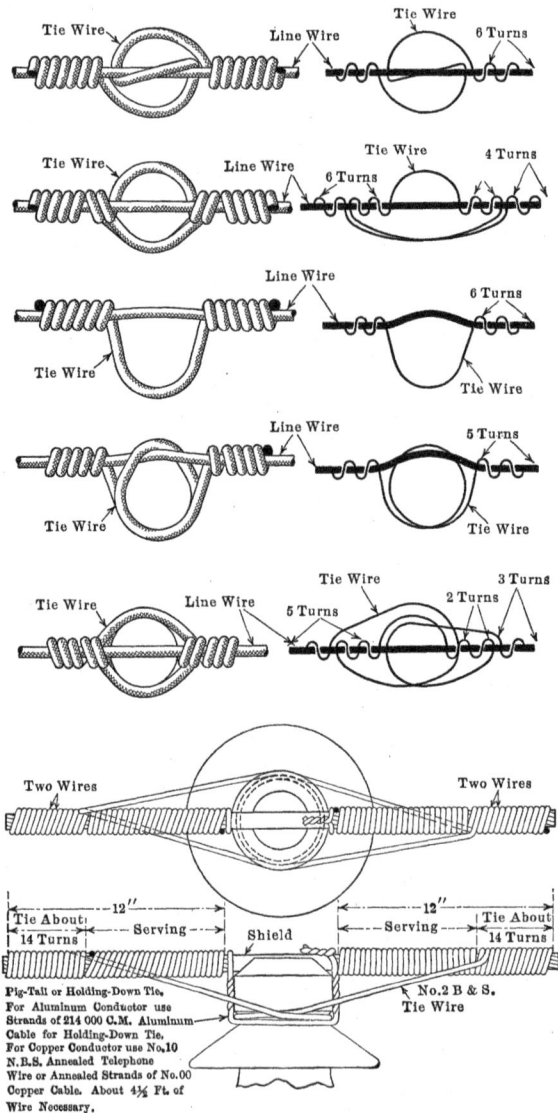

FIG. 111.—Ties.

the writer is familiar. The principle, however, applies to all cases so that, except at corners, the poles cannot be subjected to greater wire loads than the ties used will transmit. The writer does not infer that supports in general have been made too strong, but that the reasons given for the strengths used are not logical and not in accordance with the facts.

Loops.—Loop cables, either as shown in Fig. 112 or having Crosby or other clips instead of a wrapped splice, have sometimes been used to dead-end or securely attach the conductors to pin-type insulators. The loop is placed around the neck of the insulator and the end is clamped to the conductor with three-bolt clamps or several clips. If properly made, this attachment has a much greater strength than the average insulator or pin, but care must be exercised to avoid injuring the conductor. With soft-drawn cables particularly, any attempt to secure

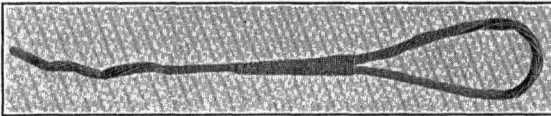

FIG. 112.—Loop cable.

strength by overtightening two clips will probably result in cutting the strands under the clips; therefore, three or more clips should be used in order to reduce the grip at any one point.

The strength of loop cables is primarily a matter of workmanship, either to secure an efficient splice or to so attach the loop to the power wire that the ultimate strength of the latter may be developed without injury to the strands.

A few tests, incidental to others, gave the following results:

Test No. 1.—A loop cable of 250,000 c.m. soft copper directly attached to the testing machine failed in the loop splice at 7200 lb., after developing the entire breaking strength of the cable.

Test No. 2.—A similar loop cable connected to the conductor with Crosby clips began to slip at 2900 lb. and continued tightening of the clips cut the strands of the cable.

Test No. 3.—A loop cable of ½-in. steel failed at 9800 lb.

Splices.—The methods of splicing conductors, ground wires and telephone wires varies considerably, the older and less efficient forms of splice being still quite common on short-span work. For the higher voltage longer span lines, however, splices are now generally made by the use of splicing sleeves, or special connectors.

By such methods, the electrical conductivity may be maintained without a very great decrease in mechanical strength. Since hard or medium-hard conductor material and high-carbon steel ground wires are commonly used in true transmission-line construction, it is not advisable to do any soldering or sharp bending at splices. The former anneals the conductor and very considerably reduces its strength while the latter is always objectionable in wires subject to heavy loading.

Sleeve splices on the other hand may be given any desired conductivity, combined with great mechanical strength. Some care is necessary in the selection of the material, length of the sleeve, and the number and character of the twists.

Even small high-strength solid steel wires sometimes used for long-span telephone wires can now be successfully spliced by steel sleeves.

FIG. 113.   FIG. 114.

Pin Insulators.—By reference to the section on insulator pins, it will be noted that few if any pin insulators are suitable for turning sharp corners, particularly with large conductors. Even providing double arms will not give factors of safety commensurate with those presumably required on the supporting construction. This weakness of the pin insulator is not, accurately speaking, chargeable to the porcelain but to the pin. To use stronger pins would require larger pin holes in the insulators and, therefore, new designs with greater neck diameter. However much one might wish that designers had foreseen the present tendency toward rational construction, the present purchaser of small quantities of insulators is compelled to turn to a different type of construction, if any considerable degree of strength is desired. The stock, or standard insulators and pins shown in all manufacturers' catalogs include a great variety of designs most of which could be dispensed with to the great advantage of both the manufacturer and purchaser. By handling fewer designs manufacturers would not be compelled to retain so many molds and so much stock, while purchasers would perforce have their choice limited to a smaller number of well-selected types. Particular reference is made to the wood top, porcelain base, wood base, or all-wood-top pins, and to all pins having bolts $\frac{7}{16}$ in. or $\frac{1}{2}$ in. in diameter. Mechanically, the excuse for such designs in their application to severe loading, defies analysis.

The strength of an insulator or of a pin should be determined by a test made after the two are assembled and with the wire and its attachment in place. Tests on pins alone may involve a reduction in the lever arm of the load, and in any case give no indication of the result of pin bending on the insulator. If there is considerable bending, which is exaggerated on wooden arms, the attachment of the wire may slip over the head of the insulator. In general it is more than possible that the values obtained by piece tests on rigid supports could not be even approximately duplicated in actual practice. That pin insulators have an entirely proper function may not be denied. Further, they are probably the most economical type for all of the more common voltages. Although used up to 80,000 volts, with the expressed satisfaction of the users, the writer is of the opinion, taking into consideration the mechanical as well as the electrical features, that their proper field of usefulness is below 60,000 volts. At heavy corners, however, the design should be changed to the disc type, in which far greater strength is obtainable. In addition it would seem entirely logical to assert *that if pin-type insulators are used on tangents and the insulators in question have only a fraction of the strength of the wire, then the supports need not be built to resist a wire tension, which the insulators can never transmit to them.*

13

There may be reasons for desiring a certain strength in the supports, but certainly broken-wire loads at maximum tension on such construction cannot be one of them.

Insulators of the suspension or disc type are more satisfactory from a mechanical viewpoint than pin insulators, since they transmit the wire tension directly to the crossarms without introducing the torsion due to the height of a pin insulator. When used either in the suspended or strain position, disc insulators require some additional height of pole or tower and an additional width of crossarm to provide clearance for the jumper or the suspended

FIG. 115.—Suspension insulators.   FIG. 116.—Through-pin insulator.

string of insulators. When used in the strain position on medium-voltage lines, it is now thought desirable to add one disc to the number used in the suspended position. For the high-voltage lines, there is now a rather general objection to using the strain position at all as it is claimed that a large proportion of the trouble from lightning has occurred at strain connections. To avoid installing insulators in the strain position and still maintain, at least in a measure, an auxiliary attachment, the arrangement shown in Fig. 162 has been used.

The types of insulators shown in Figs. 115, 116 and 117 are much stronger mechanically than single-pin or double-pin insulators, that in Fig. 116 being particularly suited for turning corners with heavy wires. It should be noted that the through pin acts as a simple beam instead of as a cantilever, and transmits its load to double arms without torsion. Further, as this pin can be made a rolled-steel rod about 1⅜ in. in diameter, it is not liable to bend and crack the porcelain. Unfortunately, however, the electrical strength of these insulators does not render them advisable for use much above 20,000 volts, although they are rated as 30,000-volt material. Their mechanical strength is usually about 12,000

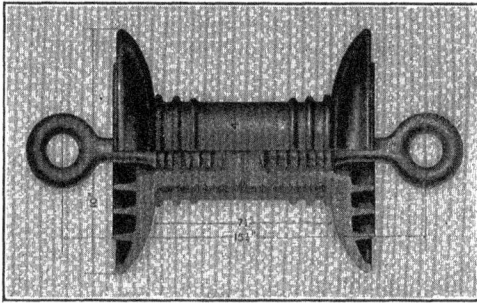

Fig. 117.—Heavy-strain insulator.

lb. A corner, particularly an important one, should be insulated approximately up to crossing requirements.

Suspension insulators, in general, have mechanical strengths ranging from 9000 lb. to 16,000 lb.

For extremely heavy stresses, strain insulators of the type designed for railway service and shown in Fig. 117 may be used. These insulators, while expensive, can be equalled in strength only by some paralleled system of the disc type, an arrangement both cumbersome and costly. The insulator shown has a mechanical strength of 20,000 lb., while the strength of a larger size is 35,000 lb. In transmission-line work these insulators have an occasional use in very long span construction.

It has sometimes been specified that the mechanical strength of guy insulators should not be less than 1, 1½, or 2 times that of the guy in which they are placed. If this requirement were strictly enforced it would penalize the use of extra heavy guys which are desirable on account of corrosion, since it would be difficult, perhaps impossible, to obtain insulators of the requisite strength. Interlocking insulators of the "goose-egg" type have great mechanical strength and retain a measure of the guy action after failure, although the guy would be slack. Such insulators, however, do not have the high electrical factor of safety of the disc type. The insulation of guys on poles carrying high-voltage wires is undesirable from a structural viewpoint, although it may at times be desirable as a matter of policy for the protection of pedestrians, particularly in the case of wooden poles.

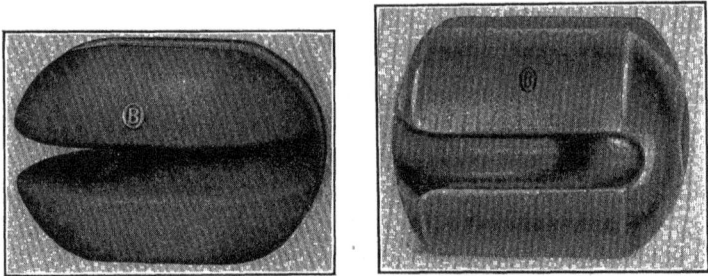

Fig. 118.—Guy insulators.

Pins.—To a certain extent, the day of the wood pin is passing, though it is still used on systems operating below 11,000 volts, and apparently gives satisfactory service. Mechanically, locust pins are inferior to metal pins and, by causing the removal of a large section of the timber, materially decrease the strength of the crossarms.

It is generally assumed that the common forms of straight-line insulator pins are "strong enough," even that locust pins will "do the work." It is true that almost any insulator and pin will support a straight unbroken line, but it is not therefore true that the same members will withstand the stresses which are in the wires under maximum loading. If the connection of the wires to the insulators will not transmit more than a fraction of the maximum stress in the wires, then the pin need withstand only the transverse load. As, however, it is often specified that the pins must dead-end the wires, it would seem necessary first to so attach the wires that they will remain attached under maximum tension and second to provide a pin and insulator which will remain intact under this stress.

Wood pins are generally of yellow or black locust and should be straight grained and free from knots except that small sound knots, ⅛ in. to ¼ in. diameter, may be permitted in locations where they will not materially impair the strength or durability of the pin. They should be free from wane or sapwood, and from checks or worm holes. The standard thread is four per inch and the threads should be cut cleanly and uniformly to provide a tight fit in the insulator. Unless well-seasoned timber is used, the pins will probably vary from the standard dimensions and the protective coatings will be less effective. Such coatings are paint, creosote, linseed oil, and perhaps more commonly paraffin.

The holes in the crossarm should be so bored that the tapered

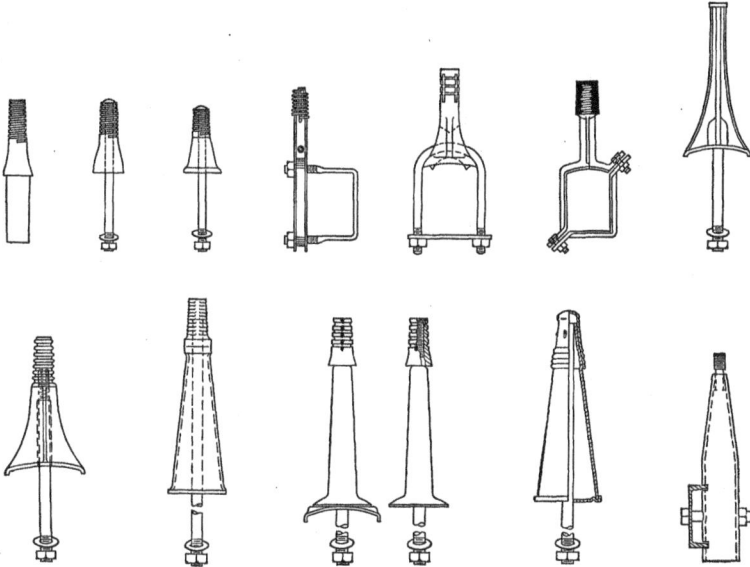

Fɪɢ. 119.—Types of insulator pins.

shanks of the pins will fit tightly therein and the pin be perpendicular. A six-penny nail may be driven into the shank from the middle of one side of the arm.

Aside from the possibility of rupturing the porcelain in case the pin bends, it has been found by test that certain classes of pins deflect as a whole and allow the top of the insulator to be subjected to shear and tension. If such pins are carried by wood arms the angular movement may be quite large, due to the penetration of the pin base in the timber, with the result that the wire fastening either slips over the head of the insulator or shears the top from the insulator.

It is probable that the character of the pin, particularly in regard to its hold on the insulator, is of equal importance with the strength of the insulator itself. A cemented pin is somewhat objectionable because it cannot be readily replaced and to overcome this difficulty various thimbles and separable-top pins have been devised.

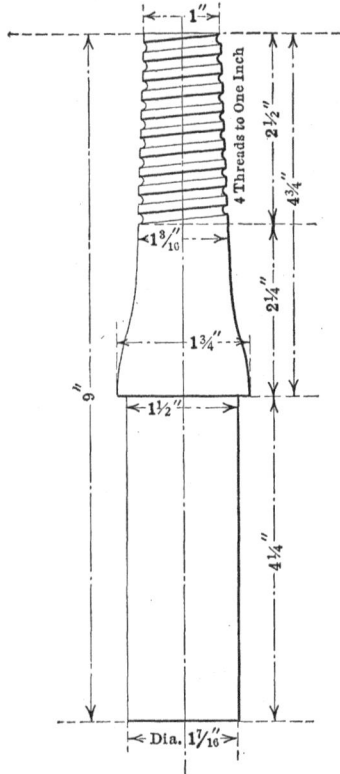

Fig. 120.—Standard locust pin.

The writer does not know of any pin having a wood thimble in which the strength of the complete pin is more than about one-half that of the bolt. This is due to the fact that the thimble is too thin and does not have sufficient bearing on the base. The weakest part of this type of pin is not in the porcelain insulator but in the design of the pin itself.

Again referring to the cemented pin and assuming it rigidly fastened to a metal crossarm, it has been found that some of the ordinary low-voltage porcelain insulators and metal pins cannot withstand long-span loading, or safely support the transverse bending caused by heavy cables on angle poles.

On high-voltage lines the insulator pins should be of metal. There are, however, a number of 11,000 to 22,000-volt lines

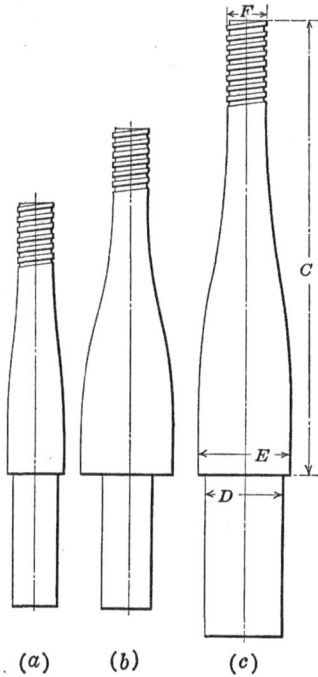

(a)   (b)   (c)

FIG. 121.—Wooden pins.

	F	C	E	D
a	1⅜″	10″	2⅛″	1¾″
b	1⅜	13	3½	2
c	1½	17	3½	3

equipped with wooden pins. The low mechanical strength of such pins and the possibility of their disintegrating or burning has raised the question of the limiting conditions under which wooden pins are permissible.

In general, pins should extend well into the insulator to reduce the mechanical stress on the material of the insulator. On account of the improbability of frequent painting, metal pins should be galvanized, or otherwise protected against corrosion.

It should be noted that in case of a broken wire some of the long pins now in use would develop a very large torsional effect upon the crossarms.

The calculated strengths of the insulator pins shown in Figs. 124, 125 and 126, are those at which the bolts should begin to bend, thereby allowing the insulators to tilt. In making the computations it was assumed that there was a complete, level contact between the pin base and crossarm and that the bolt was

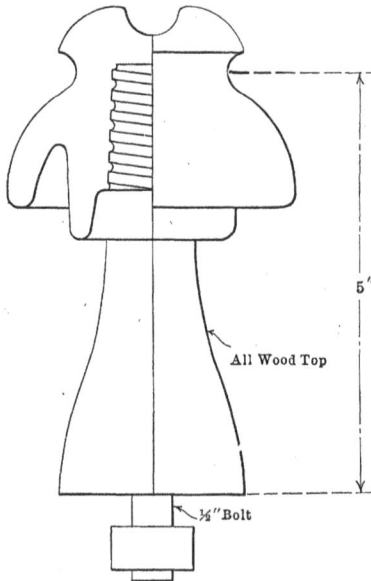

FIG. 122.—Tension applied at neck of insulator; average of three tests. 500 lb.—pin started to bend; 590 lb.—pin failed by splitting the wood thimble.

not subjected to preliminary bending by any slack in the adjustment.

If wood crossarms are used, allowing greater preliminary bending due to the bolt compressing the fibers, the strength and rigidity of the construction would be further reduced. On the other hand, if bolt steel having a yield point of 32,000 lb. were used instead of the steel of 28,000 lb. assumed above, there would be an increase of about $\frac{1}{8}$ in the tabulated strengths.

Wood-top or porcelain-base pins of this general type are all relatively weaker, owing to the fact that both timber and porcelain have a much lower crushing resistance than cast-iron, malleable-iron, or cast-steel.

The following conclusions seem to be justified: (*a*) That most of the standard designs of pins now in use are undesirable in that the metal parts are weaker than the porcelain; (*b*) that ordinary insulator pins are not at all suitable mechanically for corner construction or for dead-ending; (*c*) that much stronger pins can be designed for metal arms if a little additional thickness is allowed in the insulator neck and a larger bolt is employed.

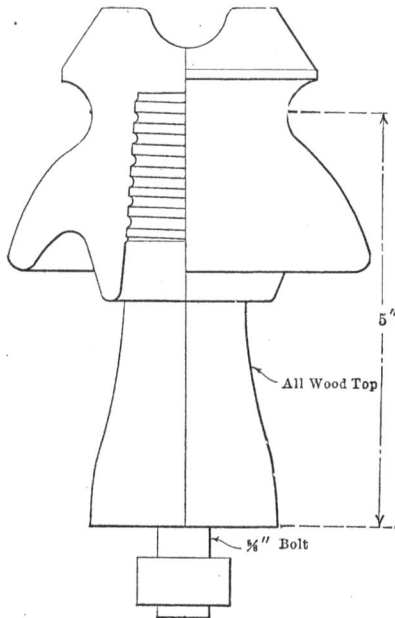

All Wood Top

5″

⅝″ Bolt

Fig. 123.—Tension applied at neck of insulator; average of three tests. 770 lb.—pin started to bend; 920 lb.—pin failed by splitting or crushing the wood thimble.

The type of pin shown in Fig. 127 was designed to avoid holes in wood crossarms. It would appear, however, that the additional cost and material are not justified, provided comparison is made with properly designed pins.

Tests have shown that under heavy loading the critical condition is often not the strength of the porcelain or of the metal pin, but the ability of the arm to resist tilting. If the arm, as a whole, will rotate under torsion, or if the base of the pin cuts into the timber, or twists on the arm, the consequent tilting of the pin may permit the wire or attachment to slip over the head of the insulator or to shear the top from the insulator. The following short series of tests indicate the foregoing tendencies:

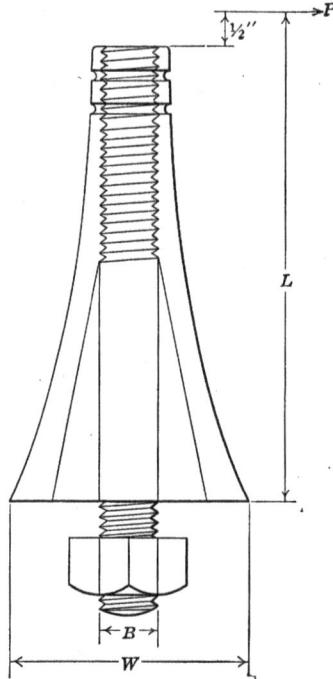

Fig. 124.—Cemented-type all-metal pin.

L	W	B	Elastic limit of bolt
5¼″	2½″	⅝″	$P = \dfrac{1.25}{5.25} \times 5600 \text{ lb.} = 1325 \text{ lb.}$
6″	2½″	⁷⁄₁₆″	$P = \dfrac{1.25}{6.0} \times 2600 \text{ lb.} = 540 \text{ lb.}$
7½″	3″	⁷⁄₁₆″	$P = \dfrac{1.5}{7.5} \times 2600 \text{ lb.} = 520 \text{ lb.}$

Tension at neck of insulator, transverse to arm.

It should be observed that the strength along the arm in the second set of tests is considerably in excess of that in the direction of the wires. In the tests in question, short yellow-pine arms were used and as standard length arms attached to a pole would have less rigidity against torsion, the effects of the tilting would have been aggravated in actual practice.

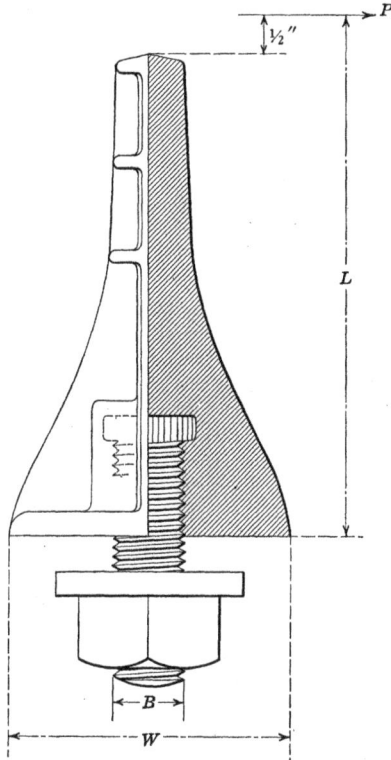

Fig. 125.—Cemented-type all-metal pin.

L	W	B	Elastic limit of bolt
5½″	3″	¾	$P = \dfrac{1.5}{5.5} \times 8450$ lb. $= 2300$ lb.
6½″	3″	¾″	$P = \dfrac{1.5}{6.5} \times 8450$ lb. $= 1950$ lb.
9½″	4″	¾″	$P = \dfrac{2.0}{9.5} \times 8450$ lb. $= 1780$ lb.

If we allow a factor of safety of 2.0 in the pin construction, and assume that the ultimate resistance to failure of some sort is about 2000 lb. for single arms and 4000 lb. for double arms,

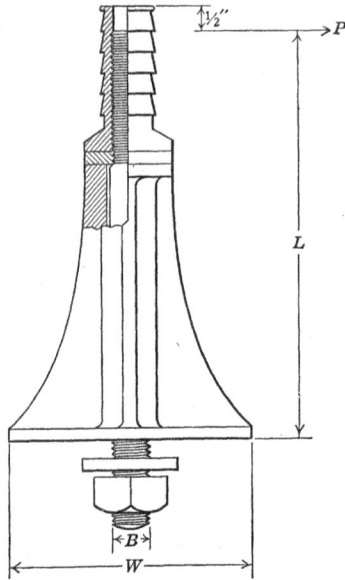

Fig. 126.—Separable-thimble type all-metal pin.

L	W	B	Elastic limit of bolt
$8\frac{1}{2}''$	$4''$	$\frac{3}{4}''$	$P = \dfrac{2.0}{8.5} \times 5480 \text{ lb.} = 1990 \text{ lb.}$
$9\frac{1}{2}''$	$4''$	$\frac{3}{4}''$	$P = \dfrac{2.0}{9.5} \times 8450 \text{ lb.} = 1780 \text{ lb.}$
$12\frac{1}{2}$	$5''$	$\frac{3}{4}''$	$P = \dfrac{2.5}{12.5} \times 8450 \text{ lb.} = 1690 \text{ lb.}$
$13\frac{1}{2}$	$5''$	$\frac{3}{4}''$	$P = \dfrac{2.5}{13.5} \times 8450 \text{ lb.} = 1560 \text{ lb.}$

Test No. 1.—2380 lb., insulator uninjured, excessive pin tilting, bottom strap bent.

Test No. 2.—2400 lb., insulator failed, excessive pin tilting, bottom strap bent. Tension at neck of insulator, along *axis* of arm.

Test No. 3.—2600 lb., insulator uninjured, pin tilted cutting into wood arm.

Test No. 4.—3300 lb., insulator failed, pin tilted cutting into wood arm, bottom strap bent and split.

Test No. 5.—4080 lb., insulator failed, pin tilted cutting into wood arm, bottom strap bent and split.

the maximum wire tensions which they would dead-end are 1000 lb. and 2000 lb., respectively. The maximum tension in

* Tests of type shown in Fig. 127.

conductors larger than No. 1 gage, however, will usually be in excess of 2000 lb.

Crossarms.—The standard crossarm, particularly for the so-called low-voltage lines, is now, and will undoubtedly remain for some years, a wood arm. Even in view of the great increase in the price of timber, the wood arm is the cheapest and the most easily obtainable throughout the country as a whole. Assuming that the price of wood arms will continue to increase, it is still probable that, for some years to come, metal or other materials will not seriously compete with timber.

FIG. 127.

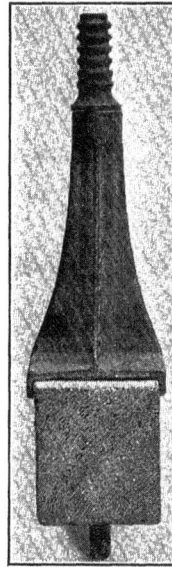

FIG. 128.

It might be supposed that preservative treatment, which will undoubtedly be extensively applied to poles, would prolong the use of wood arms. To some extent this may be true, but while the creosote treatment is rapidly growing in favor for poles, the same cannot be said of its application to arms. A preservative which would make arms less inflammable, would not drip on passersby, and would not injure the hands or clothing of workmen, would be more desirable for crossarms than creosote.

The difficulty of standardization, of foretelling accurately the uses to which an arm will be put, makes metal arms undesirable for small growing properties. The question of the relative benefits to be derived from the insulating qualities of a wood arm is a mooted one. Voltages, at least those below 13,000 volts, can on a dry day be successfully insulated by the wood arm alone.

Under such conditions, if an insulator is shattered and allows the wire to fall upon a dry or comparatively dry wood arm, no interruption of service need result nor is there any injury to wire or arm. Some unknown additional humidity, or degree of dampness of the arm, will cause burning, with the possible falling of the arm which may or may not carry other wires on the burned-off portion. The advocates of the metal arm contend that it is economical in final cost and that the insulators should be

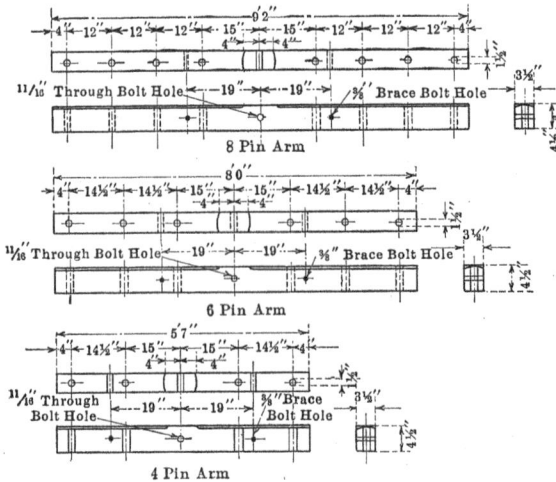

Fig. 129.—Standard wood arms.

designed to do all the insulating necessary, also that the wires falling on a metal arm shut down the service and compel proper maintenance.

The average crossarm would not withstand dead ending under maximum stress, combined with the torsional effect due to the lever arm of a long pin, without allowing a distortion which would presumably permit the wire to become unfastened from the insulators.

The more commonly used steel arms consist of single angles with the same general dimensions and punching as standard wood arms. When used on wood poles, crossarm braces are necessary.

On structural steel poles to which crossarms have two points of
connection, braces are rarely used, as it is simpler and nearly

Fɪɢ. 130.—Substantial crossarm construction for eccentric loading.

Fɪɢ. 131.—Wish-bone crossarms.

as economical in material to increase the section of the crossarms
themselves. Except with painted steel poles, and sometimes even
in that case, metal arms are galvanized.

In addition to the standard angle arms, several types of patented arms have been used to some extent, the more important being the so-called "wishbone" and the "bo-arrow" arms. In both of these, adjoining arms are brought together so that there may be two points of attachment to counteract rotation. The upper pole bolt must not be placed too near the pole top as the leverage exerted by an unbalanced pull on an arm may split the pole top. The use of special arms like these has been confined more or less to one section of the country, and it is perhaps true that familiarity will remove the sense of strangeness with which they are first seen.

Crossarm Braces.—The standard brace is a flat $1\frac{1}{4}'' \times \frac{1}{4}''$, 26 in. center to center of holes and 28 in. overall. There are also various modifications of the standard, such as changes in length and reduction of thickness to $\frac{7}{32}$ in. or $\frac{3}{16}$ in.

The term *iron* is still in common use, although the material is usually soft steel. In fact "wrought iron" or soft steel is fre-

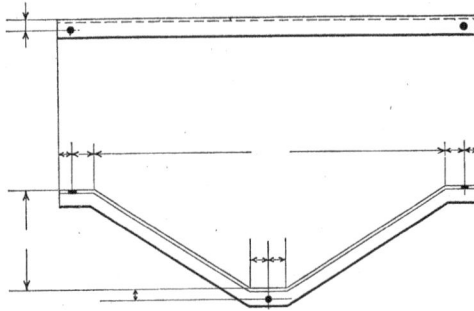

Fig. 132.—Angle brace.

quently specified, whereas a high-carbon steel would be more effective. The function of a brace is to support and prevent rotation of the crossarm, and it acts in either tension or compression. As a tension member, its strength is excessive and its rigidity against buckling as a column is, therefore, the critical condition. Owing to the shape of the section, the rigidity is a function of the thickness and the strength of the material, and as the present rather ineffective thicknesses will presumably be retained, some added stiffness may be secured by the use of the stronger steels.

The angle brace, in which two flat braces are replaced by

a single angle about $1\frac{1}{2}'' \times 1\frac{1}{2}'' \times \frac{3}{16}''$ has a much greater strength, but almost twice the weight of material at a corresponding increase in cost.

Braces have their greatest usefulness when the crossarms are unequally loaded. When but one of two circuits is installed or when all wires are placed on one side of the pole, the arms will frequently tilt, particularly on lines of medium voltages carrying heavy wires.

FIG. 133.—Crossarm with two pole bolts, no braces.

FIG. 134.—Wooden braces.

In theory at least, it is somewhat undesirable to use lag-screw connections to the pole or to the arms but, in practice, the timber about the screws will be in fair condition when replacement of the pole or a change of arms is necessary from other causes. It is probable, however, that longer, if not heavier, lag screws should be used than are always employed.

Sometimes crossarm braces have been omitted and two through-

14

bolts used to connect the arm to the pole.   This is not, however, as rigid or as strong as the more standard arrangement of one-pole bolt and an angle brace.   When a short crossarm is used for the top wire of a single-circuit pole, as shown in Fig. 133, the ground-wire post may be made a very efficient brace.

Wood braces are not good construction as it is difficult to obtain a strong permanent connection to the pole.   The pole shown in Fig. 134 is well adapted to provide adequate longitudinal pin separation, something which is not readily obtainable in double arming with large insulators.

Lag Screws or Lag Bolts.—The fetter-drive or cone-point screws generally required in former years and still shown as standard, will probably be replaced by the gimlet-point type in future work.   There is in fact no possible advantage in the former and the continuance of a double standard is quite objectionable from a manufacturing standpoint.

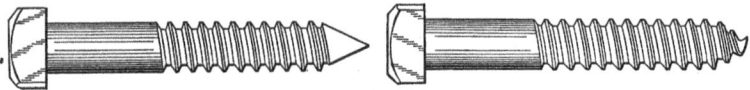

FIG. 135.—Cone-pointed lag bolt.   FIG. 136.—Gimlet-pointed lag bolt.

In theory, lag screws are supposed to be screwed into place, either into a small bored hole, or after being started by hammering.   This should be done to enable the threads to pass through the timber with the minimum shearing and injury to the fibers of the wood.

Hardware such as braces, bolts, lag screws, etc., should be galvanized by the hot-dip process, at least until such time as other methods will have clearly demonstrated an equal excellence.

It has sometimes been advocated that bolts, etc., be electrically galvanized and not subjected to the standard test, the reason being fairly obvious.

By the use of the rolled thread, in distinction to the cut thread, it is possible to use the hot-dip process without recutting the bolt threads and thereby removing the protective coating.

While not commonly done, nuts may be made with extra loose threads and after galvanizing retain at least some measure of protection on the threads.   It is probable, however, that unprotected threads on the nut and rolled threads on the bolt, both being hot-dipped, are superior to other methods.

Guys and Guying.—Guys or support braces are of three types: timber push poles, steel-cable guys and rod guys. The first are unsightly and their use is chiefly justified in places where guys are needed in two directions and can be allowed in but one. Provided a timber brace is properly set and well connected with the line pole, it is capable of resisting stresses in either direction and to some extent may act as longitudinal reinforcement. When used in such double service the setting must be adapted to resist either depression or uplift.

Under exceptional conditions rod guys may be used as they have adjustable connections, form a rigid anchorage and may be made with an excess of material to provide for corrosion. In general, however, guys are of stranded galvanized-steel cable and when properly installed are a component part of the support. The writer is utterly unable to agree with the view, sometimes expressed, that guys are a makeshift attachment of doubtful service. In fact, it is his firm belief that only the good service of the power and guy wires is retaining many existing lines in position.

It cannot be denied, however, that there are many guys of less than no usefulness owing to extreme slack, inadequate section, or excessive corrosion. A very slack guy is no guy at all, but a small, unsightly load on the structure.

So-called *iron* wire about $\frac{1}{8}$ in. in diameter is not an efficient guy, though it might serve for very light lines if the material were in reality wrought iron, and the galvanizing as efficient as would be desirable.

Steel cable not less than $\frac{5}{16}$ in., and preferably not less than $\frac{3}{8}$ in., with a heavy coat of galvanizing is the desirable standard guy. Greater strength may be obtained by the use of larger diameters, or higher grades of steel, but the latter are stiffer and more difficult to handle in the field and therefore less popular. It is possible to obtain any desired grade of steel cable from the standard-guy-strand of about 60,000 lb. per square inch to extra-high-strength steel with a strength of 180,000 lb. per square inch. In general, however, the standard or the Siemens-Martin grades should be used, but with sufficient diameter to provide ample strength, bearing in mind the much more rapid corrosion of galvanized cable than of galvanized unwiped structures.

The exact location of guys must depend on local conditions and their number on the character of the supports. With wood poles and flexible frames, guys should be used more plentifully than with semi-flexible poles, and the latter in turn require more guys than semi-rigid towers, while the true rigid towers require no guys.

In general, wood poles and flexible frames should be side guyed at all corners, at tops of steep hills, and usually wherever a very long span occurs. They should be head guyed on steep hillsides, long spans, hill tops and at intervals on long tangents.

There is no valid objection to the intelligent use of guys. Structures so designed that their light flimsy nature renders them overliable to buckling by guys should not be used in a transmission line. Further, the majority of the existing pole lines derive much of their strength from guys. *If a guy is not overtightened* its presence must inevitably increase the strength of the structure to which it is attached.

In guying it is necessary to adapt the number, size, position and tension of the guys to the service required. When practicable, guys should not be anchored too close to the structure they support. The angle of inclination, however, is not fixed.

The insertion of a turnbuckle in a guy, particularly in a long guy, permits a more careful adjustment of tension than is practicable with wire clamps.

Looping a guy around the entire tower is not good practice, except in unusual cases where there are no steel edges to be distorted and where the structure is adequately braced both vertically and horizontally at the point of guy connection. Guy connections should be made close to a panel point of the vertical bracing to prevent distortion of the main legs. When a single guy is used on the tension side of a tower it is generally desirable and frequently essential to attach it to both main legs on that side so that the pull will be exerted squarely on the tower, and not merely on one corner. Ordinarily, a guy should be attached as close as possible to the conductors whose pull it carries to the anchorage. Except in pole lines consisting of many wires where guys attached among the upper crossarms are absolutely necessary, the guys should not cross over conductors, but should be connected near the bottom crossarm.

With steel structures guy insulators are not required by the standard specifications, their use being optional, but if used they form a weak point in the guy.

T_t = total tension, or pull on pole to be balanced by guy.

T_g = total tension in guy.

$$T_g = \frac{T_t \times L}{L_g \times \sin \alpha}$$

Assuming that $L - L_g = 3$ ft., we have:

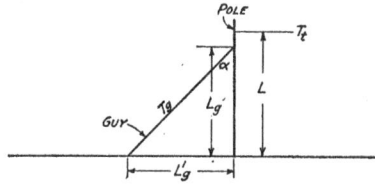

Fig. 136a.

TABLE 30.—TENSION IN GUY DUE TO T_t = 1000 LB.

L'_g (ft.)	L_g (ft.)						
	20	25	30	35	40	45	50
5	4,750	5,710	6,680	7,680	8,660	9,670	10,650
10	2,580	3,020	3,480	3,950	4,430	4,920	5,410
15	1,920	2,170	2,460	2,760	3,060	3,370	3,690
20	1,620	1,790	1,980	2,190	2,410	2,620	2,850
25	1,470	1,580	1,720	1,870	2,030	2,200	2,370
30	1,380	1,460	1,550	1,670	1,800	1,920	2,060
35	1,320	1,380	1,450	1,530	1,630	1,740	1,840
40	1,290	1,320	1,380	1,440	1,520	1,610	1,700
45	1,260	1,280	1,320	1,380	1,440	1,500	1,590
50	1,240	1,260	1,290	1,320	1,380	1,430	1,490

While guys of special steel are sometimes used to obtain great strength it may be more advisable to use heavier guys of standard material, as the latter is more easily handled and its use will tend to encourage an allowance for corrosion.

Guy Anchors.—Patented guy anchors are of many kinds, generally variations of either the old patent for screw-piles, or of the unfolding type. The diameter of the disc or blade varies from about 6 in. in the smaller sizes to 12 in. in the largest. The holding powers claimed for such devices should be used with the reservation of a factor of safety, as the character of the soil, either in general or at the time of test, exercises a very great influence on all foundation values.

The resistance of an anchor to uplift depends primarily on its depth and bearing area and on the weight and cohesion of the superimposed soil. The depth and area of the anchor blade are lim-

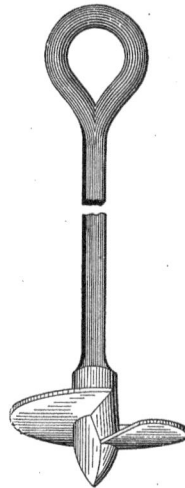

Fig. 137.

ited by the means available to install it, but the weight and cohesion of the soil will vary from place to place and from season to season.

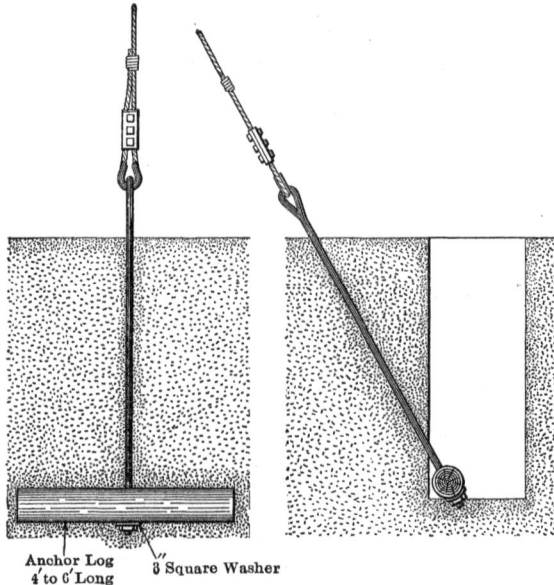

Anchor Log
4′ to 6′ Long   ⅜″ Square Washer

FIG. 138.—Wooden dead-man guy anchor.

"Dead men" are more efficient than patent anchors, since they are always larger and can be made of any desired size. Large sound logs, about 10 in. in diameter and 6 ft. long are desirable for ordinary guying. Logs treated with creosote are still better, while the best type of anchor is a concrete-covered steel beam or reinforced-concrete block. In any case the anchor rod or rods should be galvanized by the hot-dip process (N.E.L.A. Specifications) and preferably incased in concrete or in a concrete-filled pipe from the anchor to a point about 1 ft. above the ground line.

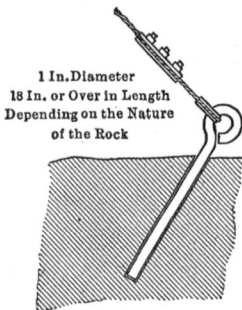

1 In. Diameter
18 In. or Over in Length
Depending on the Nature
of the Rock

FIG. 139.—Rock anchor.

The published values of the holding power of various sized anchors are apparently based on tests in clay and therefore should be reduced about 25 per cent. for sandy soil. Further, it is very difficult to arrive at any acceptable standard values for the holding power of anchors, as would be evident from an analysis of the values heretofore published. For instance, if the holding power of a 6-in. anchor buried vertically 5 ft. is 15,000 lb., the pressure on the top surface of the 6-in. disc whose area is 28.3 sq. in. would be 76,500 lb. per sq. ft., or 38 tons per sq. ft. If the holding power is due first to the weight of the cone whose sides have an inclination of 45° to the vertical (and such inclinations have sometimes been limited to 30° in foundation

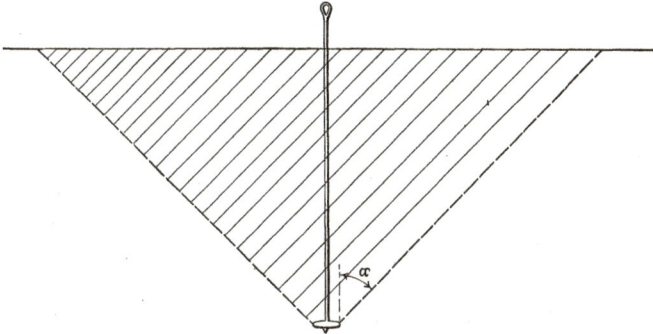

Fig. 140.

work), and second to the cohesion of the earth on the periphery of the cone, the following values will result:

Angle α assumed as 30°:

Volume of superimposed cone of earth	=	56 cu. ft.
Weight of cone at 100 lb. per cubic foot	=	5,600 lb.
Cohesion, or shear, at 150 lb. per square foot	=	9,400 lb.
Published holding power	=	15,000 lb.

Angle α assumed as 45°:

Volume of superimposed cone of earth	=	151.5 cu. ft.
Weight of cone at 100 lb. per cubic foot	=	15,150 lb.
Cohesion or shear, negligible	=	0 lb.
Published holding power	=	15,000 lb.

The pressure of 38 tons per square foot is greatly in excess of that permitted by any foundation specifications and the inclusion of cohesion is not specifically permitted by such specifications. Therefore, it would seem either that the ordinary requirements for foundations, which are generally assumed as having a factor of safety of 5 or 6, are unnecessarily conservative for anchor installations in fairly good ground, or that the generally accepted holding powers of anchors are excessive. The writer believes the former to be the case, at least under favorable conditions, but the above analysis may serve to explain the wide discrepancies in anchor values. It must be admitted that the efficiency of any given type or size of anchor will depend on the soil in which it is placed. Therefore, since the character of the soil at various anchor locations is not usually known in advance, it is, perhaps, advisable to use disc or unfolding anchors only for light guys and to rely on the installation of "dead-men" to resist heavy stresses. Far greater holding power can be obtained by the use of a good dead-man than can possibly be provided by any of the patented anchors whose area is necessarily much less than that of any dead-man. Since the initial stress on the guy anchor will be approximately one-third of its maximum stress, care should be taken to disturb as little of the adjoining earth as possible during construction, in order that the anchor may have a high initial resistance without depending on the additional strength resulting from a future compacting of the soil.

CHAPTER XII

ERECTION AND COSTS

Erection.—In stringing wires it is, of course, of importance to cover as much ground daily as possible, but this should not be done at the expense of injury to such an important and expensive item of the construction as the wire. Copper wires, whether solid or stranded, cannot be dragged without injury over the ground or over the crossarms. If either of these methods is adopted, it will result in nicks in the solid wire, or broken strands in the cables. Such injuries may not be visible and, with good fortune, may never cause failure, but anyone who has seen soft stranded copper wire snarl into a veritable "rat's nest" when *removed* through snatch blocks, will not deny that injuries may result from improper stringing.

It is also necessary to be constantly on the lookout to avoid kinks, twists, or broken strands, either in unreeling or in stringing. When broken strands or injurious kinks do occur, a new section of cable should be spliced into the line. It is almost impossible to remove the cable from a reel without forming kinks unless the reel rotates about its axis. The reel should, therefore, be supported on a horizontal shaft arranged to turn freely, but not too fast, as all cable has a tendency to kink as soon as a little slack occurs. If kinks do form it is, of course, desirable to remove them if possible to do so without injury to the cable. By immediate attention and adherence to the proper methods of manipulation, it is possible to remove a kink without injury to the strands. To do this it is necessary, as shown in Figs.[1] 141, 1, 2, 3 and 4, to straighten the wire by pushing the ends apart without altering the lay of the strands. If this is not done correctly, some of the strands will be stretched and the spiral of the wires will be distorted at the point of bend, and thereafter the cable will fail by the parting of the individual strands at much less than the ordinary strength of the cable.

In stringing wires they should be pulled out through snatch

[1] Illustrations from Yellow Strand—Broderick and Bascom Rope Co.

blocks with wooden sheave and frame and ball or special bearings and afterward lifted into place on the insulators or in the clamps. Any simple device which can be quickly attached to a

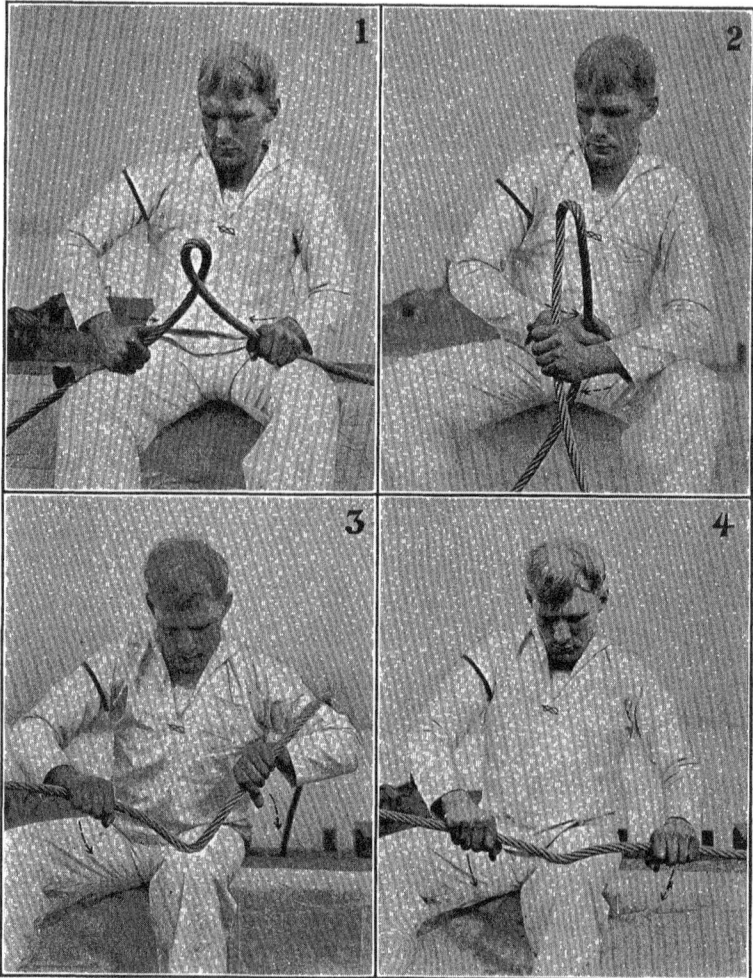

Fig. 141.—Straightening kinks.

crossarm to hold the top groove of the snatch block at the elevation of the clamp, will be found to be very useful. Periodic inspections should be made of the condition of the snatch blocks, to prevent injury to the wire. If a dynamometer is used to ad-

FIG. 142.—Stringing wire with derrick car.

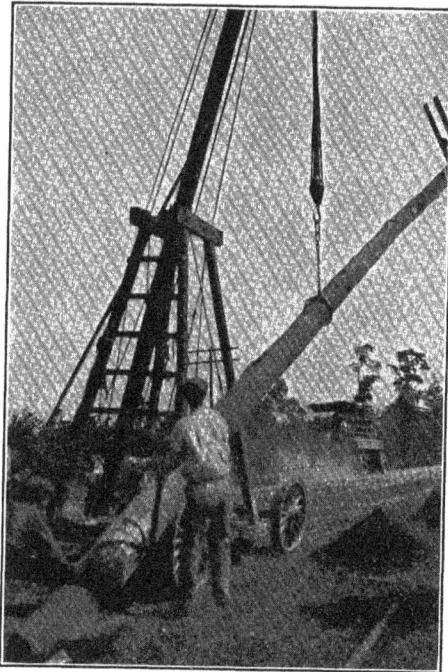

FIG. 143.—Derrick wagon raising pole.

just the sags the wires can be transferred from the blocks to their final positions.　Dynamometer stringing is particularly desirable in long-span construction, although it is uncommon in the more general classes of short-span work.　It should be checked occasionally by measurement of the resulting sag.　It is advisable to string all spans to balance at normal temperature and no wind or ice, even though this results in some unbalancing under load. Otherwise, the tensions might balance at maximum load, perhaps once in 10 years, and be unbalanced the rest of the time, with the consequent continuous loading of their supports.

When a track is available paralleling the pole line, erection with a derrick car will give the greatest possible distances per day. In the absence of a track it is frequently practicable to use motor or horse-driven derrick cars for the erection of poles or for string-

FIG. 144.—Derrick wagon.[1]

ing wires.　The economy of such erection equipment is very great under favorable conditions, but it must be kept moving to attain its greatest usefulness.　For this reason no direct comparison can be made between the erection costs of two lines, if one is long and accessible, and the other is short and inaccessible.

The wagon derrick, shown in Figs. 143 and 144, consists of a wide stout wagon base carrying short, double shear legs from which is hung a wood mast.　The mast is hung from an axle and a universal joint, thus allowing the top to travel in an arc limited by the length of the groove which restrains the base.　By this arrangement sufficient overhang is obtained to reach holes about 10 ft. from the wagon.　The mast is back guyed by ropes attached to the top ring and snubbed around any convenient object, or bar driven in the ground beyond the wagon.　Since

[1] Designed and built by Wm. A. Ladue, Supt. Public Service Elec. Co., Hoboken, N. J.

the height of the shear legs is not great and as the mast may be rotated in both directions, the rig with the mast in the horizontal and axial position can pass under bridges, trolley wires, etc.

When but one of two ultimate circuits is to be immediately installed, as shown in Fig. 145, it is frequently placed on one side of the pole and on the highway or most accessible side. This practice is not commendable as it tends to tilt the arms, as is faintly visible in the illustration, and because erection of the second circuit is made much more difficult. When substantial angle braces are used such eccentric loading is probably unob-

FIG. 145.—Highway-side loading.

jectionable under ordinary conditions, but it has been observed that all side-arm construction suffers more in severe storms than balanced arms. If, however, only standard flat braces are used, some arms are certain to become tilted.

Placing the first wires on the highway side of a pole line reduces slightly, but only slightly, the original cost of erection. The second circuit, however, will have to be erected under much less favorable conditions than the first would have been if placed on the inside. It is better construction, when both circuits are to occupy the top arm, to place two wires on the inside and one on the outside. This will necessitate longer shut-downs on the first circuit during the second stringing, but it will reduce the ultimate cost of that stringing and provide a stronger original line.

Poles should be set vertically and in line, except that at corners and dead-ends they may be given a slight rake, though this is unusual for steel poles. In building foundations or setting poles, particularly in long-span lines, it is necessary to exercise some care to obtain a firm unyielding support for the pole. Since providing a layer of concrete in the bottom of the excavation is not always practicable, or might result in considerable expense, the bottom of the excavation should be compacted by tamping, and perhaps by adding and tamping a 6-in. layer of broken stone or gravel. The back filling should always be well tamped in thin layers, and to insure this it is frequently required that one shoveler be used to three tampers. If broken stone or gravel is removed from the

Fig. 146.—A frame erection with gin-pole.

excavation, or readily obtainable nearby, a very efficient foundation may be obtained by back filling with a considerable proportion of rock, being careful to pack earth or sand in the spaces between the pieces. Since the back filling will settle and become more compact after it has been completed, and after rains, a portion, at least, of the excess excavation should be piled up around the base of the pole. Later an examination should be made of the foundations and back filling added wherever it may have settled below the surface.

On wood-pole lines it is customary to set adjacent poles with the crossarm gains facing opposite ends of the line. Guys should be installed before any wires are strung and should be inspected and adjusted if necessary after the stringing is completed, otherwise, the structures may receive an overloading, while without the guys.

If some wood poles have unusually large tops, the regular crossarm bolts may be too short.   In such cases it is better to obtain a few long bolts, than to injure and weaken the top of the pole by cutting it down to the shorter bolts.   An occasional heavy pole, or one with a large top and regular taper, is a real asset to the structural strength of any wood-pole line, therefore such poles should not be weakened by excessive top cutting.

Costs.—In most contract work it is fairly accurate to assume that in general the work will approach the estimated average

Fig. 147.—Erection with house-derrick.

and that under-estimates of some portions of the work will be balanced by exceptional records made in other portions where organization and familiarity are given a fair test.   For this to be true, however, it is necessary that there be a fairly large volume of work in a few locations.   Where work is scattered the expense is in moving, beginning, and stopping, not in the actual work itself. Further, there are more *kinds* of work in line construction than in most other classes of contract work, since each pole or tower location is a small job and in some way unlike the last.   It is, therefore, impossible to attain the speed of piece-work in a fixed location, as a very considerable portion of an employee's time is spent in "thinking" about the next step, or in moving to a new position, or in getting a new tool.

In a comparatively short installation, even so unconsidered an item as a specially rainy week will have a marked effect on the unit cost per structure. Rain or snow not only stops or delays progress—except the progress of the "straight-time" pay-roll—on the day in question, but also usually delays the work of the following day. Holes are filled with water or snow, equipment or material is buried, slides have occurred, walking and teaming is more difficult, and in general it is a poor day's rain which cannot count as two.

Fig. 148.—Hauling a small concrete pole.

Work carried on between spring and late fall should cost at least 20 per cent. less for labor than that done during the remainder of the year.

Instances are rare in which published accounts cover the matter of accident insurance. An owning company may and perhaps usually does, include such contingencies in overhead expense, but the cost is nevertheless directly applicable to the line erection. A contracting concern, on the other hand, usually tabulates the insurance as part of the cost estimate and as it is almost universally paid as a direct percentage on the actual labor pay-rolls, it is a very real item. The amount paid for insurance varies with different classes of labor and in different states. Under certain recent legislative enactments, the liability of employers is now only partly protected by premiums as high as 15 per cent., so the casual omission of such items of expense is at least censurable.

Again in comparing the costs of a previously established method of construction for any company, such as the ordinary wood-pole line, with for example, steel-pole lines, the omission of general expense items in the former is quite common. For instance, the company may maintain a small force of travelling inspectors and purchasing agents in order to obtain their quota of wood poles, and this expense with that of handling, storage, trimming, etc., is properly chargeable to the cost per pole delivered. In general, the work done at odd times on regular construction mate-

Fig. 149.—Raising a small concrete pole.

rial will not be charged thereto as certainly as the unusual charges to new types of construction.

Proper charges for plant and equipment, particularly for small tools, are noticeable by their absence in most cost estimates.

It has been stated that the cost of steel poles or towers varies directly as the square of the height, a very evident error since their weight, under constant conditions of design, will be more nearly in direct proportion to the height, and the greater heights usually indicate a smaller number of structures per mile. As a matter of fact, it is surprising to note the relatively small difference in estimated cost between different designs of equal or nearly equal general excellence. The conditions of manufacture, accessibility of the site and character of the ground usually influence the ultimate cost much more than is often realized. In addition,

15

it will be found by investigation that very few existing lines are directly comparable on account of differences in design.

Published accounts of erection costs are usually misleading and frequently inaccurate. Unless the local conditions are similar and the methods of erection equally efficient there will be no equality between two sets of costs. Furthermore, a difference in the extent of the work and in the organization of the field forces may cause a relatively great difference in the cost of two lines of exactly similar construction.

In making a comparative cost estimate of two different types of construction, as for example wood poles and steel poles, it is necessary either to make two complete and distinct tabulations, or to use care to include all credits and debits due to the differences in construction. In either case an accurate estimate should include allowances for maintenance, renewals, and interest charges.

In making estimates it should be remembered that the use of a long-span steel-pole line will effect a saving of about two-thirds of the cost of the insulators, pins, ties, pole rights, and foundations, and of the erection of the insulators, pins and ties. Having fewer insulators than the shorter span wood pole line, there will be less probability of insulator failure and, therefore, less interruption to service. In addition, some credit is probably due long-span steel construction on the ground that the maintenance expense will be lower and that a high-grade line with few poles will cause less criticism.

If shop-assembled steel poles with galvanized butts approximately 24 in. square are employed, they can be set in the ground without concreting the holes, so the cost of the foundations per pole should not greatly exceed that for wood poles, and a saving of nearly two-thirds of the hole digging would result.

Steel poles are distributed and erected the same as wood poles, therefore with the proper field equipment a mile of steel poles should be set at least as quickly as a mile of wood poles.

Established costs of concrete-pole lines are practically non-existent, and even such pole costs as have been published are rarely applicable to ordinary transmission line work. The more recent, best built, and most important concrete-pole lines have been for telephone and telegraph, rather than power service. Some short "back-yard" poles for purely distribution service have been built at an approximate average cost of $10 per pole. True transmission line poles of adequate height and strength, unless made in quantities, would probably cost at present, about the same as structural steel poles.

As previously stated, it is necessary to use considerable judgment in making cost estimates or in interpreting them. The following costs which have been compiled from time to time, as well as the author's estimates, cannot be considered universally applicable. Indeed they are only reasonably accurate for particular cases.

CALGARY WOOD-POLE LINE

(*Electrical World*, Jan., 1912)

One ground wire, ¼-in. steel.
Three conductors, No. 0 aluminum, 55,000 volts.
Two telephone wires.
Pin insulators.
Spans, 150 ft.
Height, 40-ft. poles.
Average cost per mile, $2000.

* * * STEEL-POLE LINE

One ground wire, ⁷⁄₁₆-in. steel.
Three conductors, No. 2 copper, 33,000 volts.
Two telephone wires, No. 10 copper-clad.
Span, 400 ft.
Height of poles, 43 ft.

Poles...............................	$690	
Freight and cartage..................	25	
Foundations.........................	65	
Guying..............................	30	
Erection............................	55	
		$865
Wires...............................	$685	
Insulators, pins and ties.............	145	
Erection............................	105	
		$935
Clearing, damages, etc...............	25	
Right-of-way........................	195	
Supervision.........................	100	
Miscellaneous.......................	110	
Total cost.......................		$2230

* * Wide-base Tower Line

This line was erected under adverse weather conditions, and in extremely rough, and rather inaccessible country. The soil was hard clay.

One ground wire, No. 00 copper.
Six conductors, No. 00 copper.
(One circuit installed).
Standard spans, 800 ft.

	Cost[1] per tower
Hauling	$14.50
Setting footings	73.20
Assembling tower	24.80
Raising	22.00
	$134.50

Hauling—four-horse teams, driver and one to three helpers.
    —a portion of line material distributed from railroad.
Setting —Foreman
    —four diggers
    —six to eight templet, level and survey men.
Assembling—foreman and 5 to 20 men, depending on weather.
Raising—foreman, four-horse teams, shear legs, etc., and eight men (average four towers per day).
Cost of common labor, $2.25 a day.

[1] Average of 104 towers.

* Steel-pole Line (Narrow-base Towers)

This line was erected in 1905, under circumstances that were not at all favorable to a low cost. The work was done in winter weather; the foundations were expensive both in design and in construction, and the line is crooked and difficult of access

Twenty-four conductors, 250,000 circ. mils, soft-drawn stranded copper, 11,000 volts.
Eight feeders, 500,000 circ. mils, soft-drawn stranded copper, 650 volts.
Wood crossarms.
Standard spans, 150 ft.
Pole heights, 40, 45, 50, and 55 ft.

Weight of standard 40-ft. pole	3000 lb.
Weight of standard 45-ft. pole	3300 lb.
Weight of standard 50-ft. pole	3800 lb.
Weight of standard 55-ft. pole	4000 lb.
Weight of special or angle poles	5000 lb. to 6000 lb.

APPROXIMATE AVERAGE COSTS

Poles	Cost per mile
Steel poles	$3,550
Wood crossarms	200
Erection	550
Foundations	5,800
Guying	50
Painting poles and arms	200
Total	$10,350

ESTIMATED COST. ONE-CIRCUIT LINE WITH STEEL POLES

One ground wire.
Three conductors, No. 1 copper, 22,000 volts.
Two telephone wires.
Standard spans, 450 ft
Poles per mile, 12.

Poles:

		Cost per mile	
Material	at $50 per pole	$600	
Freight		35	
Hauling	at $2.50 per pole	30	
Foundations (earth)	7 at $2 = $14		
Foundations (braced)	3 at $3 = $ 9		
Foundations (concrete)	2 at $11 = $22		
	12 $45	45	
Erection	at $3 per pole	35	
Guying		15	
Painting		20	
		$780	$780

Wires and Line Material:

	Cost per mile	
One ground wire ⅜-in. galvanized steel cable	$60	
Three conductors, No. 1 stranded copper	670	
Two telephone, No. 6 BWG Siemens-Martin steel	80	
	$810	
Ties, guys, splices, etc	50	
33,000-volt insulators and pins		
27 insulators on tangent poles		
18 insulators on corner poles		
45 insulators and pins at $0.75 each =	$35	
45 telephone insulators and pins at $0.20 each =	$10	
Hauling	10	
Erecting	110	
	$1025	$1025
Clearing, trimming, etc		25
Right-of-way at $5 per pole		60
Supervision		100
Contingencies and miscellaneous		25
Total per mile of standard line		$2015
Crossings and special structures		$

The following comparative estimates of the costs of a steel-pole line with long-span construction, and of a wood-pole line with short-span construction, both for the same location in accessible rolling country, indicate an ultimate saving in favor of the steel line.

ESTIMATED COST. ONE-CIRCUIT LINE WITH WOOD POLES

One ground wire, ⅜-in. steel.
Three conductors, No. 2 copper, 33,000 volts.
Two telephone wires, No. 10 copper-clad.
Standard spans, 120 ft.
Poles per mile, 44.
Pin-type insulators.
Metal arms.

Poles:

		Cost per mile	
Poles 35 ft. long, 7-in. tops............at $5 each		$220	
Crossarms, galvanized........................		167	
Telephone brackets......................		5	
Pole steps and hardware..... ..at $0.75 per pole		33	
Framing and trimming........at $0.50 per pole		22	
Creosoting butts..............at $0.20 per pole		9	
		$456	
Hauling.......................at $1 per pole		$ 44	
Digging holes.........at $1.20 each $ 53			
Bog shoes or braces............... 6			
Setting poles.........at $1.80 each 79			
Miscellaneous.................... 4			
$142		$142	
Guying.......................................		30	
		$672	$672

Wires and Line Material:

One ground wire................... $ 54			
Three conductors................. 544			
Two telephone 50			
$648			
Ties............................. 5			
Soldering materials................ 5			
33,000-volt insulators.............. 66			
Pins............................. 49			
Telephone insulators.............. 5			
Ground-wire connection............ 16			
Stringing 3 miles, No. 2 copper...... 45			
Stringing 2 miles, No. 10 copper-clad.. 20			
Stringing 1 mile, ⅜-in. steel........ 18			
Miscellaneous 4			
$881			$881

Clearing, trimming, etc.........................	10
Miscellaneous materials and tools	15
Right-of-way....................at $5 per pole	220
Supervision, engineering and general expense....	100
Contingencies and miscellaneous	25
Total per mile of standard line.............	$1923
Crossings and special structures.............	$

ESTIMATED COST. ONE-CIRCUIT LINE WITH STEEL POLES

Wires same as before, except that a $\frac{7}{16}$-in. steel ground wire was assumed.

Standard spans, 400 ft.

Poles per mile, 13.

Three-disc suspension-type insulators.

Poles:

		Cost per mile	
Material.......................at $53 per pole		$689	
Hauling....................at 2.25 per pole		29	
Digging holes.........at $1.50 each $19.50			
Concrete at corners................40.00			
Crushed stone.....................6.00			
	$65.50	$65	
Erection.....................at $2.25 per pole		29	
Guying..................................		30	
Painting.....................at $1.50 per pole		20	
Miscellaneous................................		8	
		$870	$870

Wires and Line Material:

One ground wire...................	$75		
Three conductors..................	544		
Two telephone wires..............	50		
	$669		
Soldering materials, etc............	5		
Insulators and clamps.............	137		
Telephone insulators..............	5		
Stringing 3 miles, No. 2 copper	54		
Stringing 2 miles, No. 10 copper-clad.	24		
Stringing 1 mile, $\frac{7}{16}$-in. steel	20		
Miscellaneous...................	6		
	$920		$920

Clearing, trimming, etc........................	10
Miscellaneous materials and tools......... ...	20
Right-of-way........ .... at $7 per pole	90
Supervision, engineering and general expense...	100
Contingencies and miscellaneous.............	25
Total per mile of standard line.............	$2035
Crossings and special structures............	$

CHAPTER XIII

PROTECTION

Ground Wires.—In the light of our present knowledge,"ground" or "sky" wires seem to be desirable on lines of 11,000 volts or more, but are not necessary on lines which are in more or less sheltered locations. If, however, the ground wire is of less durable material than the conductors under it, or is improperly connected

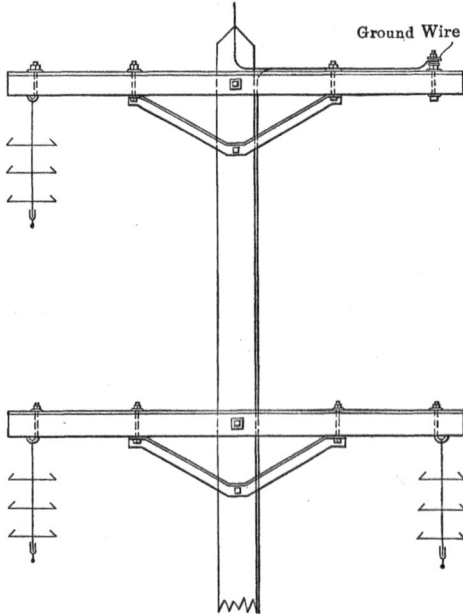

Ground Wire

Fig. 150.—Eccentric location of ground wire.

to the supports, it becomes a menace rather than a safeguard. A poorly constructed ground wire will eventually cause interruptions in the service of the power wires below it.

The relative merits of galvanized-steel, galvanized-iron, copper-covered and copper wire are not definitely known and the subject is worthy of much more careful consideration than it has thus far received. The second material mentioned, *i.e.*, galvanized iron, is more or less of a misnomer, as there is said to be little or no real iron wire used for transmission purposes. A large portion of the so-called iron wire is in reality soft steel, which does not have the ability to resist corrosion like the old-fashioned wrought-iron wire.

In the process of galvanizing wire cables the excess coating is wiped off, resulting in a thinner coat than is usually obtained on galvanized shapes which are merely allowed to drain. This explains in some measure the increased life of windmill towers over that of guy cables. It need not be inferred that galvanized ground wires are undesirable in all instances, as in certain localities they will prove economical. In general, however, the probable life of galvanized wire in the locality under consideration should be scrutinized with care before such material is placed immediately over copper conductors.

The choice between copper-covered and copper cable, is chiefly a matter of cost, if it is assumed that the relatively thin shell of the copper-covered cable will be effective in preventing corrosion. A smaller gage copper covered steel wire or cable

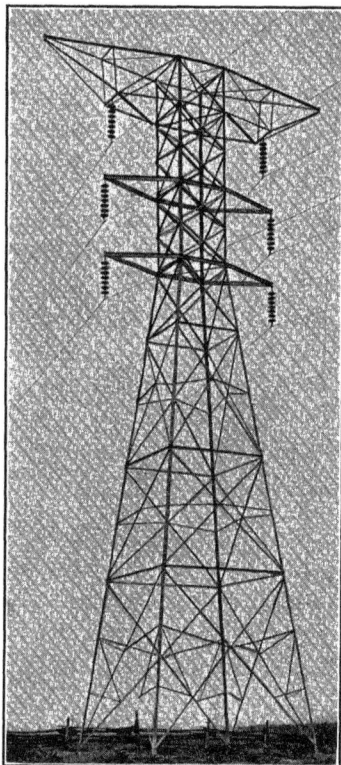

FIG. 151.—Two-circuit tower, two ground wires.

will have the strength necessary to permit a sag equal to that of the larger copper cables. If a copper-covered ground wire is used, it should have a heavy coating of copper.

A number of lines have been built with ground wires of the same section and material as the conductors, but the greater number have galvanized-steel cable. It is also probable that the majority of ground wires are of a smaller gage than the power wires. Good practice seems to indicate, however, that galvanized-steel ground wires should be cables $\frac{3}{8}$ in. or more in diameter and that copper-covered stranded-steel should . be heavily coated. Furthermore, it is desirable to use cables having few strands in order to obtain a relatively thicker coating of copper.

The ground wire connection differs from the power connections in that it should be treated as a dead-end connection at every pole or tower. A variation in sag due to the accidental, or intentional, slip of a conductor has less opportunity to cause trouble than a similar slip in the ground wire. In the case of the so-called flexible towers, by which is meant those having little theoretical strength in the direction of the line, a firmly attached ground wire is needed to serve as a partial guy to help minimize the extent of tower failure. The ground wire attachments should, therefore, be well tightened, regardless of the condition of the power-wire attachments.

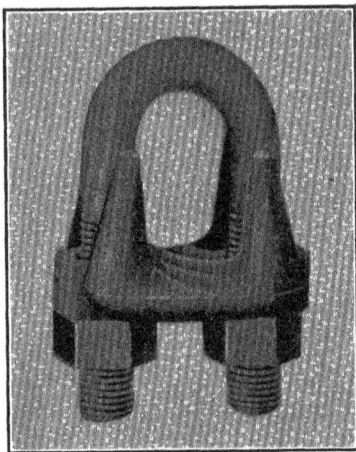

FIG. 152.—Crosby clip.

On some supporting structures the ground wire is connected with a vertical earth wire leading to a ground plate beneath the support. Such connections should be arranged to preserve, as much as possible, the original strength of the ground wire. The junction is necessarily at the point of maximum mechanical stress in the ground wire; therefore soldered or bent connections are particularly undesirable.

The proper location for a ground wire with respect to the conductors is at the apex of a 60° to 90° angle enclosing the latter wires. In recent practice the ground wire is usually placed a distance above the upper conductors equal to about one-half the horizontal space between them. On some high-voltage lines with two circuits in vertical spacing, two ground wires have been used, one over each set of conductors. This method undoubtedly gives some increased protection, but its relative effectiveness is uncertain.

On the other hand some designs for one-circuit poles have placed the ground wire in the position opposite the upper insulator connection, so that it is in no sense over two of the power wires (Fig. 150).

It seems, therefore, in considering the protection afforded by overhead ground wires and in judging the results obtained in actual installations, that some weight should be given to the relative location of wires on the lines in question.

The attachment of the ground wire to its supporting structure

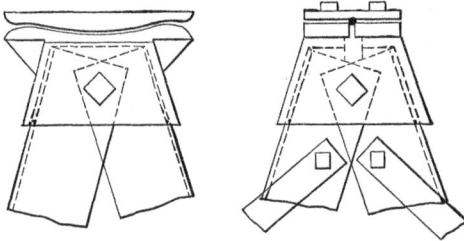

Fig. 153.—Ground wire clamping cap.

is a detail of great importance. If a long, smooth, well-rounded wire seat in the clamps is a wise provision for the attachment of the power wires, it is equally desirable for a ground wire which overbuilds the power wires. It is a matter of common knowledge that a short rigid metallic connection with a small U- or hook-bolt biting into a wire has a tendency to cause wire failure. Therefore, such connections should not be used in the very worst possible place on a power line.

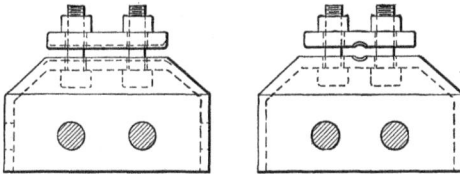

Fig. 154.—Ground wire clamping cap.

Further, the use of ridges or teeth in the contact surface is decidedly poor practice in any connection to copper wire. Such projections merely serve as cutting edges to injure the softer copper material when the device is tightened. The indications are that better results are obtained with copper wires by the use of a long contact surface without change of direction, rather than by short clamps with pronounced waves. Since clamps are frequently galvanized, care should be taken to obtain a smooth surface finish, both before and after galvanizing, on the portions which will come into contact with the wire. Sand spots or edges on the black material, or improper draining of the zinc coating, may cause the formation of sharp projections which will injure the wire.

The ends of the wire grooves should be bell-mouthed with a gradual slope, in order that the wire may not be bent sharply, or even appreciably, at any point. In other words, there should be a reasonably long tangent contact surface ending in curved orifices, the sides of which should confine the wire, as nearly as possible, to the exact position where its curve of sag would cause it to lie under any conditions of loading.

Neighboring Lines.—Whether or not the location of a proposed transmission line is in a measure fixed by existing property rights, as may be the case with lines on electric railways, it is necessary, or at least very advisable, to confer with the owners of any existing and adjacent wire lines to predetermine what measures, if any, may be required to prevent interference with the proper operation of such foreign lines.

There are two general classes of neighboring lines; other transmission lines, and the so-called "no-voltage" lines such as the telephone and telegraph.

In relation to both of these classes of lines, there are two classes of adjacency, crossings and parallelism. Crossings may vary from single-span, right-angle crossings to several-span, oblique or "skew" crossings. The term parallelism is ordinarily used to indicate two lines on separate structures, though it also applies to over building whether on the same or separate structures.

Interferences may also be divided into two classes, inductive and contact. Interferences of the first class are probably confined to cases of parallelism while the latter may occur with either parallel or crossing lines.

The theory of inductive interference is not yet thoroughly understood, so generally effective measures for its prevention have yet to be devised. It appears that induced currents may occur with rather widely separated lines, between which there can be no physical contact either direct or indirect. The matter of induction, therefore, should be considered as a separate subject. In its relation to the physical characteristics of a transmission line, induction need, therefore, be considered only as a reason for the inclusion of transpositions.

Interferences by contact may be of many kinds and may occur wherever two lines are near each other. Reasonable security at crossings is not difficult to obtain, but the matter becomes rather complicated where transmission lines and no-voltage lines are closely parallel.

In general, it will be found advantageous to locate the proposed power line as far as practicable from the no-voltage line, preferably on a different route. Otherwise, the two lines should be separated as widely as possible, and ordinarily occupy opposite sides of the highway or right-of-way. In some instances it will be necessary to move all, or parts, of an existing line in order to maintain a proper separation. Again it may be advisable to consolidate two existing lines to provide space for the power line.

In view of the fact that most of the existing lines are of relatively remote origin, and together with the power line frequently subject to governmental supervision, it should be unnecessary to point out the propriety of an investigation as to the relative rights, contracts and responsibilities of the conflicting lines. Unfortunately such investigations seem to have been the exception rather than the rule.

In an attempt to reduce the possibilities of interference by contact, it immediately appears that physical separation is the most effective method. The amount, or distance, of separation is not a fixed quantity, nor is it essentially a function of any physical characteristic of either line. Each installation should be considered as a special case since the local topographical conditions have almost as much bearing on the effective separation as the pole heights, spans, factors of safety and details of construction of the lines. For example, it is evident that there can be no physical contact between the supporting structures of two lines, if they are separated by a high embankment.

Again, a low line cannot come into structural contact with a tall line if the poles of the short line are set opposite or near the mid-span of the other. In both of the above cases, however, there is a possibility that wind-blown wires of either class may afford contact with the parallel line. The relative positions of the two lines with regard to locations on side hills and exposure to storms will have a very direct bearing on the possibility of contact. In addition to the foregoing, consideration must be given to the presence of inflammable material near either line, the probability of wind-blown branches, etc., and the details of construction of both lines.

An investigation, in 1914, of the number of failures at crossings in the States of Idaho, Oregon, and Washington on a total of 1953 crossings, for periods from 2 to 8 years, showed a total of six failures, only one of which resulted in the damage to the company crossed.[1]

Voltages	Crossing years	Failures	Damage
5,000– 7,000	635	0	0
11,000–15,000	3,397	4	None
22,000–44,000	1,209	0	0
55,000–66,000	5,544	2	Phones burned ($25.00)
	10,785	6	

Cradles.—A cradle or guard net is a wire basket—in the more elaborate form, a wire tunnel—formerly used to separate two power systems, or to protect a telephone, telegraph, or similar system from an electric light or power line. A cradle to be effective must practically inclose one system of wires, since there is no justification for the assumption that a broken wire will fall vertically and remain within the confines of a flat net of restricted area.

That the use of cradles was a natural development in the progress of transmission line construction, may be admitted; it is, however, an indisputable fact that in recent years they have fallen into great disfavor.

Inasmuch as prevention is more desirable than cure, the latest practice in this country is to so install the power line that the conductors will rarely break and in some instances to further insure against a falling wire by the use of an auxiliary attachment designed to hold the wire in case of the failure of the insulator or of the wire at the insulator.

Clamping Devices.—The Joint Report specifications for crossings (Edition of 1911) state explicitly what the connection of the power wires to the supporting structures at crossings should accomplish, but do not state the exact means by which the result should be attained.

The general theory of the clamping device is to require an effi-

[1] Idaho Power and Light Mens Assoc.

·cient insulator and a positive dead-ending attachment of the wire
to the insulator through a device having sufficient mechanical
strength to resist the tension of the wire in case the insulator fails;
that the device should have sufficient mass to resist burning, at
least to some extent; and that the points of attachment to the
power wires be at a sufficient distance from the insulator to minimize the danger from arcs. The attachment should be so de-

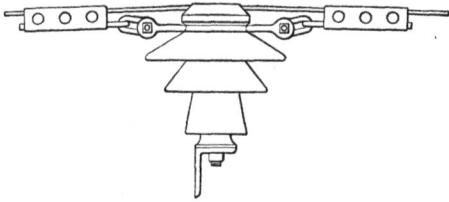

Fig. 155.

signed that it cannot fall free of the pin in case the insulator is
shattered. It should not require delicate adjustment and
should be firmly clamped to the wire without injuring it in any
way.

In order to prevent the burning of wood pins and crossarms by
arcs from defective insulators, or fallen wires, by causing the
circuit breakers to act, some specifications have required that

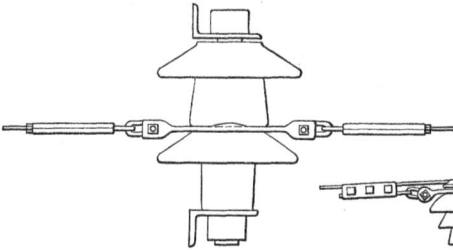
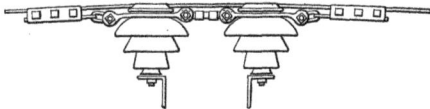

Fig. 156.   Fig. 157.

crossarms should be of metal, or be provided with metallic strips
and that they should be grounded. In other instances, metal
grounding arms have been placed below the wires so that a falling
wire would come into contact with them, as in the H-frame crossing, Fig. 45, in which the auxiliary chain attachments have not
yet been clamped to the power wires. On the other hand, the
grounding of wood crossarms results in the loss of the insulating

value of the wood arm. In dry weather a wire falling on a wood arm would not necessarily burn either the arm or the wire.

With insulated wire there is a possibility of injuring the wire in removing the insulation and this may outweigh the effect of the insulation in preventing a tight grip on the power wire itself.

It is not clear to the writer how a device at the support can insure against accidents from breaks out in the span. Moreover, it would appear that the majority of failures occur at the insulators, so that the greatest practicable benefit would be obtained by improving, if any improvement is necessary, the construction at the insulators. It is claimed by some engineers that nothing is required except a first-class insulator and a tie wire, and that the cost of auxiliary devices would better be expended for higher grade insulators. Further, as the low-voltage lines are not subject to much electrical trouble, a slight increase in insu-

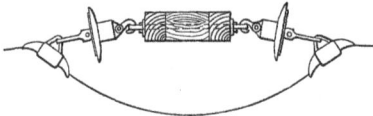 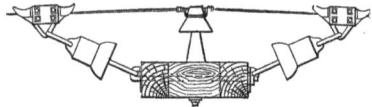

Fig. 158. Fig. 159.

lation might practically insure such lines against failure at the insulators.

In some cases an auxiliary or second attachment of the power wire is used, but unfortunately this is not always as effective as it appears. For instance, if such an arrangement requires dead-end connection on a single-pin insulator, it is in itself undesirable and mechanically impracticable for heavy stresses. It has also been proposed that the wire be protected from arcs by the use of an arcing strip, cap, etc., but these are not effective if a shattered insulator allows the attachment and wire to fall. Again, it is not possible by the use of any device at the support to prevent a wire broken out in a span from coming into contact with a line beneath it. However, as the majority of failures occur at the insulators, it seems wise to neglect this possibility and concentrate attention on other features of the connections at the supports.

In the writer's opinion too little value has been attached to the ability of a wire to hold, over a doubled span length, in case of pole failure. With ordinary short-span construction and reasonably low voltages there is little reason for doubting that the wires will be less liable to injury if the crossarms are ungrounded. Therefore, any grounding device should include some provision to prevent the actual separation of the wire into two spans, either of which may fall.

The clamping device or dead-ending attachment, illustrated in Fig. 160, is typical of the erroneous idea that protection may be afforded foreign interests without due regard to the protection of the power line. In other words, the benefits from the method

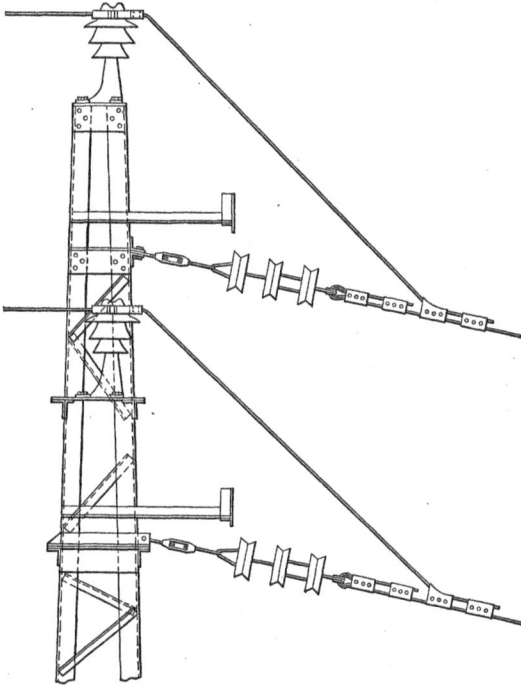

Fig. 160.

used to eliminate danger from falling conductors are more than offset by the possibilities of accident introduced by the protective construction.

As shown, the span adjoining the crossing is dead-ended on one pin type insulator and this insulator must carry a heavy mechanical load, not merely when a break occurs, but at all times. Apart from the fact that this is one of the things most to be avoided in line material installation it introduces an unbalanced load on the support. In general, it compels the insulators, pins

16

and towers to withstand a continuous loading far in excess of
their continuous loading under ordinary construction.

The crossing span is supported from the strain insulator connection and the power cable is not continuous over the supports,
the clamps having to serve both as a mechanical and an electrical
connection.   In actual construction it is probable that the power
cable would be passed around a large stout thimble and possibly
"served" at the end connection; otherwise the detail seems an
undesirable one for copper cable.   The constant mechanical
stress to which the strain insulators are subjected is undesirable
and at high voltages might be very objectionable, though it is true
that such insulators have greater mechanical strength than those
of the pin type.

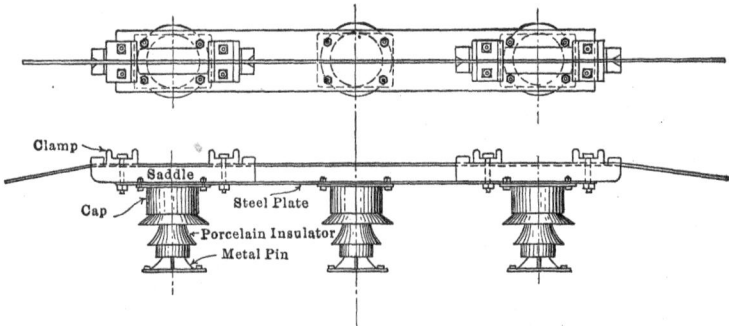

FIG. 161.—Pin insulators with caps and saddle.

Further, it seems at least possible that an electrical breakdown
of the strain insulators might cause an arc which would melt the
jumper cables immediately above them thus rendering the auxiliary connection useless.

The purpose of the grounding arm beneath the jumper is to
ground the latter in case a conductor breaks outside of the clamps,
i.e., out in the span.   There does not appear to be any great
assurance, however, that the ground would be effective before
the long end of the wire falling free could come into contact with
wires beneath it.

The use of the rigid clamping cap on the pin insulators is not
considered the best practice, the general tendency now being to
allow the power cables to balance themselves about a smooth
porcelain surface, and to have any auxiliary connections "ride"
on the line.

If there is any basis of fact in the recent theory that lightning troubles are aggravated by bends in conductors, this construction would be open to such criticism.

Nests of insulators supporting a cast saddle have been used to some extent for the conductors of unusually long spans, such as river crossings and in some instances for railroad crossings (Fig. 161). It is possible in this manner to provide sufficient pin strength to withstand heavy stresses and at the same time obtain a long contact surface to which the conductors can be clamped. The chief objection to this construction, apart from the cost, is the difficulty of removing and replacing shattered insulators. There is usually a considerable weight of cable carried by the

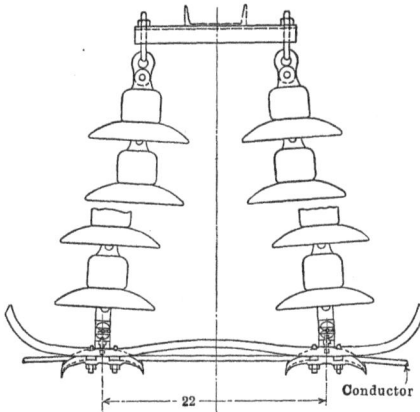

Fig. 162.—Two strings of insulators, suspended position.

saddle and also a vertical component of the wire tension, due to the fact that the wires often descend from the saddle to a lower adjoining pole. The clamping attachment of the saddle as well as all portions of the wire seat should be smooth and well rounded to avoid breakages due to the rigidity of the saddle and the vibration of the wires.

Another and perhaps a better method of supporting moderate-length spans is to use two strings of suspension type insulators hung as shown in Fig. 162. Two strings of insulators in the strain position have also been used and have generally given satisfactory service except at the higher voltages.

The design shown in Fig. 162 has the merit of a lower mechanical stress in the insulators, and should be less conducive to impact in case of wire failure while retaining, in a measure, the auxiliary or double connection. This method of attachment, however, requires considerable separation between crossarms to provide clearance for wires which descend sharply on one side of the tower.

Several years' experience with a number of unusually high river crossing towers in different localities has led the writer to believe in the propriety of over-insulating the wire attachments on such towers. Considering six pairs of towers ranging in height from 96 ft. to 188 ft. separated from 10 to 100 miles and all in a region subject to rather severe electrical storms, there have been but two cases of insulator failure. Both insulators were of the single-disc type and were hung in the strain position. While the insulators were badly burnt and shattered, in neither case did they break apart or allow the wire to fall. Two crossings have pin insulators and the others have disc insulators in the strain position. The failures occurred on one of the two sets which did not have an overhead ground wire. One failure resulted directly from lightning and the other apparently from a heavy surge following a head puncture, by lightning, of a pin insulator on another structure. All of the towers are grounded. No other trouble of any nature has been reported on any other wires, although the various structures carry from six to twenty-four 13,000-volt wires each.

DISCUSSION OF JOINT REPORT

SPECIFICATIONS FOR CROSSINGS

(Edition of 1911–1914)

The writer was a member of two committees signing the Joint Report and perhaps for the very reason that he has been compelled to occupy a double standpoint some explanatory discussion of various disputed requirements may not be out of place. The fact is frequently forgotten that this specification is a pioneer and in its very nature a compromise. There was previously no standard even approximately acceptable to conflicting interests, whereas with the Joint Report as a basis, an electric service company and any company whose lines are crossed are able to agree promptly on mutually satisfactory construction without the old interminable delay.

Objections have been made that the specifications, when literally enforced, are oppressive in certain special cases. This is not a reasonable objection since it may be said with equal truth of any specification.

To consider the specifications by clauses we have:

1. SCOPE.—The limitation of 5000 volts in relation to telephone lines, while no doubt desirable from a telepone standpoint, may be a hardship to power lines located in streets. If it is a fact that protective apparatus is not effective above 2300 volts, there seems little logic in limiting the power line to 5000 volts, if the probability of failure on a well-constructed 13,000-volt is not measurably greater than that of the ordinary 2300-volt line. The limitation is an old standard which conceivably should not now apply with equal force under improved methods of construction.

The words "constructed over" are perhaps unfortunate since they introduce the inference that the clause applies to joint poles and parallel lines. In reality the specification is as stated a *crossing* specification, a crossing being a single span or perhaps two or three spans.

8. CLEARANCE.—Side clearances of (12 ft. 0 in.) are not properly enforceable in case the track occupies a city street within that distance of the curb, provided always that reasonable clearance may be obtained, or that trees, etc., already occupy the curb line.

10. While the 8 ft. 0 in. clearance above other wires is reduced by paragraphs 16 and 33, to 6 ft. 0 in. under certain conditions, it would be an improvement to specifically permit 6 ft. 0 in. clearance for all construction in which the wires are securely fastened and when the poles are not subjected to much bending. Further, such clearances should be extended to cover the voltages above those embraced in the various clearances given in the specifications.

12. Dead-ending through disc insulators in the strain position should not be mandatory as more recent experience tends to prove this method undesirable in some instances. Provided other requirements are complied with, this clause should not be blindly enforced.

19. GUYS.—While inferentially guys are permitted on any type of structure, they should be specifically permitted with proper regulation.

23. GROUNDING.—As indicated heretofore, the writer, personally, is not in favor of a general enforcement of the grounded arm and in view of the general lack of agreement on the subject since developed by the specification, it would seem preferable to make the requirement voluntary.

34. LOADS.—The sliding scale of broken wires, while apparently a reasonable compromise between none broken and all broken, should specifically include provision for designing with "pullback" from the unbroken wires. This is what actually occurs in fact on heavy lines and unless this interpretation is made, such lines are unduly punished. Further, it is illogical to design all multiple-pin arms for large broken cables without pullback. Pullback is not mentioned directly in the specification, though it may be inferred.

37. FACTORS OF SAFETY.—The pin and insulator are a structural unit after erection and the porcelain does not really need a larger factor

than the pin, nor is its individual factor readily determinable. For heavy cables the factors given for pins and insulators are prohibitory if literally enforced.

The factor of 3.0 for structural steel is not exactly correct as a single statement and should be omitted since the matter is covered by the allowable unit stresses given in paragraph 69.

Concrete poles if properly made are unnecessarily penalized if a factor of 4.0 is used.

The factor of 2.0 for foundations should not be blindly enforced as the conditions of loading presuppose at least a semi-frozen soil and the general methods of computing foundations err on the side of safety.

45. CONDUCTORS.—There seems to be little justice in using this paragraph to require an existing solid wire on ordinary construction to be replaced by a stranded cable. The clause is founded on the greater strength of large stranded cables for long-span construction and on the greater chance of injury to solid wires in erection. Therefore, short spans and careful erection, particularly with insulated wire, would be unjustly treated by a literal and general enforcement.

48. INSULATORS.—The strength of the guy insulator should be twice that of the guy stress, not of the guy strength. This is evidently a confusion of intent since the latter would penalize extra strong guys used for protection against corrosion.

Again, the interlocking feature is not generally applicable, indeed impracticable for high voltages and the requirement should not be mandatory.

54. CONCRETE.—At the time the specifications were written, the then forthcoming Joint Committee Report on Concrete and Reinforced Concrete seemed the most general authority on the subject. There is now grave question, however, that it can be applied to crossings, since it covers a different field of work, *i.e.*, bridges and buildings, and is only fairly enforceable in certain individual requirements, though educationally of benefit.

59. STRUCTURAL STEEL.—Reference to the preceding sections on structural design will show that the use of large ratios of l/r are not conducive to the best types of construction, nor safely permissible for general use. Some reduction in the figures given, such as to 150 and 200, would be no great hardship and of some real benefit to the general excellence of future work.

60. Similar reasoning indicates an increase in minimum thickness to $\frac{3}{16}$ in. for galvanized material.

64. FOUNDATIONS.—Although accurate data on foundation design are wanting, the strict application of the clause works a hardship by inferentially omitting the unknown value of earth shear, arch action, etc.

73. TIMBER.—Paragraphs 37 and 73 seem unnecessarily conservative, particularly for selected timber treated with preservative, or whose

deterioration will be closely watched. It may be granted that a fair and accurate requirement is almost impossible, but some increase of unit stress is certainly reasonable for lines in city streets.

In conclusion and in spite of the foregoing interpretations many of which have been developed only by actual use, the writer again repeats his firm belief that the specifications, if used with an honest desire to correlate divergent interests, are a very efficient work and a great advance on all previous measures.

CHAPTER XIV

JOINT REPORT SPECIFICATIONS FOR OVERHEAD CROSSINGS OF ELECTRIC LIGHT AND POWER LINES

(Edition of 1911–1914)

GENERAL REQUIREMENTS

1. *Scope.*—This specification shall apply to overhead electric light and power line crossings (except trolley contact wires), over railroad right-of-way, tracks, or lines of wires; and, further, these specifications shall apply to overhead electric light and power wires of over 5000 volts constant potential, crossing telephone, telegraph or other similar lines. It is not intended that these specifications shall apply to crossings over individual twisted pair drop wires, or other circuits of minor importance where equally effective protection may be secured more economically by other methods of construction.

2. *Location.*—The poles, or towers, supporting the crossing span preferably shall be outside the railroad company's right-of-way.

3. Unusually long crossing spans shall be avoided wherever practicable and the difference in length of the crossing and adjoining spans generally shall be not more than 50 per cent. of the length of the crossing span.

4. The poles, or towers, shall be located as far as practicable from inflammable material or structures.

5. The poles, or towers, supporting the crossing span, and the adjoining span on each side, preferably shall be in a straight line.

6. The wires, or cables, shall cross over telegraph, telephone and similar wires wherever practicable.

7. Cradles, or overhead bridges, shall not be used beneath the crossing wires or cables; but in cases where the crossing wires or cables cross beneath the railroad wires, telephone, telegraph, or other similar wires, a protection of adequate strength and proper design between the two sets of crossing wires or cables may be required.

8. Unless physical conditions or municipal requirements prevent, the side clearance shall be not less than twelve (12) ft. from the nearest track rail, except that at sidings a clearance of not less than seven (7) ft. may be allowed. At loading sidings sufficient space shall be left for a driveway.

9. The clear headroom shall be not less than thirty (30) ft. above

248 .

the top of rail under the most unfavorable condition of temperature and loading. For constant potential, direct-current circuits, not exceeding 750 volts, when paralleled by trolley contact wires, the clear headroom need not exceed twenty-five (25) ft.

10. The clearance of alternating-current circuits above any existing wires, under the most unfavorable condition of temperature and loading, shall be not less than eight (8) ft. wherever possible. For constant potential, direct-current circuits, not exceeding 750 volts, the minimum clearance above telegraph, telephone, and similar wires may be two (2) ft. with insulated wires and four (4) ft. with bare wires.

11. The separation of conductors carrying alternating current supported by pin insulators, for spans not exceeding 150 ft., shall be not less than:

Line voltage	Separation
Not exceeding 7,000 volts	12 in.
Exceeding 7,000, but not exceeding 14,000	20 in.
Exceeding 14,000, but not exceeding 27,000	30 in.
Exceeding 27,000, but not exceeding 35,000	36 in.
Exceeding 35,000, but not exceeding 47,000	45 in.
Exceeding 47,000, but not exceeding 70,000	60 in.

For spans exceeding 150 ft. the pin spacing should be increased, depending upon the length of the span and the sag of the conductors.[1]

With constant potential, direct-current circuits not exceeding 750 volts, the minimum spacing shall be ten (10) in.

12. When supported by insulators of the disc or suspension type, the crossing span and the next adjoining spans shall be dead-ended at the poles, or towers, supporting the crossing span, so that at these poles, or towers, the insulators shall be used as strain insulators, or the height of the wire attachments shall be such that with the maximum sag in the crossing span, occurring from failure of the construction outside the crossing span, and taking into account the deflections in the strings of suspension insulators, the minimum clearances, as given in Paragraphs 9 and 10, shall be maintained.

13. The clearance in any direction between the conductors nearest the pole, or tower and the pole, or tower, shall be not less than:

Line voltage	Clearances
Not exceeding 10,000 volts	9 in.
Exceeding 10,000, but not exceeding 14,000	12 in.
Exceeding 14,000, but not exceeding 27,000	15 in.
Exceeding 27,000, but not exceeding 35,000	18 in.
Exceeding 35,000, but not exceeding 47,000	21 in.
Exceeding 47,000, but not exceeding 70,000	24 in.

[1] NOTE.—This requirement does not apply to wires of the same phase or polarity between which there is no difference of potential.

14. *Conductors.*—The normal mechanical tension in the conductors generally shall be the same in the crossing span and in the adjoining span on each side.

15. The conductors shall not be spliced in the crossing span nor in the adjoining span on either side.

Taps to conductors in the crossing span are generally objectionable, and should not be made unless necessary.

16. The ties or devices for supporting the conductors at the poles, or towers, shall be such as to hold the wires, under maximum loading, to the supporting structures, in case of shattered insulators, or wires broken or burned at an insulator, without allowing an amount of slip which would materially reduce the clearance specified in Paragraphs 9 and 10.

17. *Ground wires* when installed as protection against lightning, shall be thoroughly grounded at each of the crossing supports. In case of their installation on steel supporting structures, they may be clamped thereto. In case they are installed on wooden structures, the ground wire shall be grounded at each of the structures with a solid copper wire, with as few bends as possible, and no sharp bends, and not less than No. 4 B. & S. gage or equivalent copper section. The ground wire itself, in the crossing span and the adjacent spans, may be of the same material as the conductors, or a steel strand not less than 5/16 in. in diameter may be used, double galvanized, and having a breaking strength of not less than 4500 lb. and in general shall follow the minimum factors of safety as provided for the rest of the crossing construction.

If crossarms are grounded, the same ground wire may be used for grounding the lightning protection wire as in grounding crossarm strips.

18. Where there is an upward stress at the point of conductor attachment, the attachment shall be of such type as to properly hold the conductor in place.

19. *Guys.*—Wooden poles supporting the crossing span shall be side-guyed in both directions, if practicable, and be head-guyed away from the crossing span, and the next adjoining poles shall be head-guyed toward the crossing span. Braces may be used instead of guys.

20. *Strain insulators* shall be used in guys from wooden poles, except when the guys are through grounded to permanently damp earth.

The insulators shall be placed not less than eight (8) ft. from the ground. Strain insulators shall not be used in guying steel poles or structures.

21. *Clearing.*—The space around the poles, or towers, shall be kept free from inflammable material, underbrush and grass.

22. *Signs.*—In the case of railroad crossings, if required by the railroad company, warning signs of an approved design shall be placed on all poles and towers located on the railroad company's right-of-way.

23. *Grounding.*—For voltages over 5000 volts, wooden crossarms, if used, shall be provided with a grounded metallic plate on top of the arm which shall be not less than ⅛ in. in thickness and which shall have a sectional area and conductivity not less than that of the line conductor. Metal pins shall be electrically connected to this ground. Metal poles and metal arms on wooden poles shall be grounded.

24. The electrical conductivity of the ground conductor shall be adjusted to the short-circuit current capacity of the system at the crossing and shall be not less than that of a No. 4 B. & S. gage copper wire.

25. *Temperature.*—In the computation of stresses and clearances and in erection, provision shall be made for a variation in temperature from − 20°F. to + 120°F. A suitable modification in the temperature requirements shall be made for regions in which the above limits would not fairly represent the extreme range of temperature.

26. *Inspection.*—If required by contract, all material and workmanship shall be subject to the inspection of the company crossed; provided that reasonable notice of the intention to make shop inspection shall be given by such company. Defective material shall be rejected and shall be removed and replaced with suitable material.

27. On the completion of the work, all false work, plant and rubbish incident to the construction shall be removed promptly and the site left unobstructed and clean.

28. *Drawings.*—If required, by contract, ...
(......) complete sets of general and detail drawings shall be furnished for approval before any construction is commenced.

LOADS

29. The conductors shall be considered as uniformly loaded throughout their length, with a load equal to the resultant of the dead load plus the weight of a layer of ice ½ in. in thickness and a wind pressure of 8 lb. per square foot on the ice-covered diameter, at a temperature of 0°F.

30. The weight of ice shall be assumed as 57 lb. per cubic foot (0.033 lb. per cubic inch).

31. Insulators, pins and conductor attachments shall be designed to withstand the mechanical tension in the conductors under the maximum loadings with the designated factor of safety.

32. Sags should be such that the stress on the pin falls within the limits of paragraph 31, unless methods be employed to prevent an undue slip in case of pin failure. (See paragraphs 9, 10 and 16.)

33. The pole, or towers, shall be designed to withstand, with the designated factor of safety, the combined stresses from their own weight, the wind pressure on the pole, or tower and the above wire

loading on the crossing span and the next adjoining span on each side. The wind pressure on the poles, or towers, shall be assumed at 13 lb. per square foot on the projected area of solid or closed structures and one and one-half times the projected area of latticed structures.

34. The poles, or towers, shall also be designed to withstand the loads specified in paragraph 33, combined with the unbalanced tension of:

2 broken wires for poles, or towers, carrying 5 wires or less.
3 broken wires for poles, or towers, carrying 6 to 10 wires.
4 broken wires for poles, or towers, carrying 11 or more wires.

35. Crossarms shall be designed to withstand the loading specified in paragraph 33, combined with the unbalanced tension of one wire broken at the pin farthest from the pole.

36. The poles, or towers, may be permitted a reasonable deflection under the specified loading, provided that such deflection does not reduce the clearance specified in paragraph 10 more than twenty-five (25) per cent. or produce stresses in excess of those specified in paragraphs 69 to 73.

FACTORS OF SAFETY

37. The ultimate unit stress divided by the allowable unit stress shall be not less than the following:

Wires and cables...................................... 2
Pins... 2
Insulators, conductor attachments, guys............... 3
Wooden poles and crossarms............................ 6
Structural steel...................................... 3
Reinforced-concrete poles and crossarms............... 4
Foundations... 2

NOTE.—The use of treated wooden poles and crossarms is recommended. The treatment of wooden poles and crossarms should be by thorough impregnation with preservative by either closed or open-tank process. For poles, except in the case of yellow pine the treatment need not extend higher than a point 2 ft. above the ground line.

38. *Insulators.*—Insulators for line voltages of less than 9000 shall not flash over at four times the normal working voltage, under a precipitation of water of ⅙ in. per minute, at an inclination of 45° to the axis of the insulator.

39. Each separate part of a built-up insulator, for line voltages over 9000, shall be subjected to the dry flash-over test of that part for five consecutive minutes.

40. Each assembled and cemented insulator shall be subjected to its dry flash-over test for five consecutive minutes.

The dry flash-over test shall be not less than:

Line voltage	Test voltage
Exceeding 9,000 but not exceeding 14,000....	65,000
Exceeding 14,000 but not exceeding 27,000....	100,000
Exceeding 27,000 but not exceeding 35,000....	125,000
Exceeding 35,000 but not exceeding 47,000....	150,000
Exceeding 47,000 but not exceeding 60,000....	180,000
Exceeding 60,000.........................	3 times line voltage

Each insulator shall further be so designed that, with excessive potential, failure will first occur by flash-over and not by puncture.

41. Each assembled insulator shall be subjected to a wet flash-over test, under a precipitation of water of $\frac{1}{8}$ in. per minute, at an inclination of 45° to the axis of the insulator.

The wet flash-over test shall be not less than:

Line voltage	Test voltage
Exceeding 9,000 but not exceeding 14,000....	40,000
Exceeding 14,000 but not exceeding 27,000....	60,000
Exceeding 27,000 but not exceeding 35,000....	80,000
Exceeding 35,000 but not exceeding 47,000....	100,000
Exceeding 47,000 but not exceeding 60,000....	120,000
Exceeding 60,000.........................	twice the line voltage

42. Test voltages above 35,000 volts shall be determined by the A.I.E.E. Standard Spark-gap Method.

43. Test voltages below 35,000 volts shall be determined by transformer ratio.

MATERIAL

44. *Conductors.*—The conductors shall be of copper, aluminum, or other non-corrodible material, except that in exceptionally long spans, where the required mechanical strength cannot be obtained with the above materials, galvanized or copper-covered steel strand may be used.

45. For voltages not exceeding 750 volts, solid or stranded conductors may be used up to and including 0000 in size; above 0000 in size, stranded conductors shall be used. For voltages exceeding 750 volts and not exceeding 5000 volts, solid or stranded conductors may be used up to and including 00 in size; above 00 in size, conductors shall be stranded. For voltages exceeding 5000 volts, all conductors shall be stranded. Aluminum conductors for all voltages and sizes shall be stranded.

The minimum size of conductors shall be as follows:

No. 6 B. & S. gage copper for voltages not exceeding 5000 volts.

No. 4 B. & S. gage copper for voltages exceeding 5000 volts.

No. 1 B. & S. gage aluminum for all voltages.

46. *Insulators.*—Insulators shall be of porcelain for voltages exceeding 5000 volts.

47. For pin type insulators, there shall be a bearing contact between the pin and the insulator pin hole up to the level of the top of the tie wire groove, the purpose being that the pin should directly take the strain imposed upon the insulator.

48. Strain insulators for guys shall have an ultimate strength of not less than twice that of the guy in which placed. Strain insulators shall be so constructed that the guy wires holding the insulator in position will interlock in case of the failure of the insulator.

For less than 5000 volts, strain insulators for guys shall not flash over at four times the maximum line voltage under a precipitation of water of one-fifth of an inch ($\frac{1}{5}$ in.) per minute, at an inclination of 45° to the axis of the insulator. For voltages of more than 5000 volts, the strain insulator or series of strain insulators shall not fail at the line voltage under the above precipitation conditions.

49. *Pins.*—For voltages of 5000 and over, insulator pins shall be of steel, wrought iron, malleable iron, or other approved metal or alloy, and shall be galvanized, or otherwise protected from corrosion. (See paragraph 47.)

50. *Guys.*—Guys shall be galvanized or copper-covered stranded steel cable not less than $\frac{5}{16}$ in. in diameter, or galvanized rolled rods, neither to have an ultimate tensile strength of less than 4500 lb.

51. Guys to the ground shall connect to a galvanized anchor rod, extending at least 1 ft. above the ground level.

52. The detail of the anchorage shall be definitely shown upon the plans.

53. *Wooden Poles.*—Wooden poles shall be of selected timber, reasonably straight, peeled, free from defects which would decrease their strength or durability, not less than 8 in. in diameter at the top, and meeting the requirements as specified in paragraphs 19, 33, 34 and 37.

54. *Concrete.*—All concrete and concrete material shall be in accordance with the requirements of the Report of the Joint Committee on Concrete and Reinforced Concrete.[1]

STRUCTURAL STEEL

55. Structural steel shall be in accordance with the Manufacturers' Standard Specifications.

[1] NOTE.—This may be found in the February, 1913, Proceedings of the American Society of Civil Engineers, Vol. 59, No. 2, pp. 117–168.

56. The design and workmanship shall be strictly in accordance with first-class practice.

57. The form of the frame shall be such that the stresses may be computed with reasonable accuracy, or the strength shall be determined by actual test.

58. The sections used shall permit inspection, cleaning and painting, and shall be free from pockets in which water or dirt can collect.

59. The length of a main compression member shall not exceed 180 times its least radius of gyration. The length of a secondary compression member shall not exceed 220 times its least radius of gyration.

60. The minimum thickness of metal in galvanized structures shall be $\frac{1}{4}$ in. for main members and $\frac{1}{8}$ in. for secondary members. The minimum thickness of painted material shall be $\frac{1}{4}$ in.

Protective Coatings

61. All structural steel shall be thoroughly cleaned at the shop and be galvanized, or given one coat of approved paint.

62. *Painted Materials.*—All contact surfaces shall be given one coat of paint before assembling.

All painted structural steel shall be given two field coats of an approved paint.

The surface of the metal shall be thoroughly cleaned of all dirt, grease, scale, etc., before painting and no painting shall be done in freezing or rainy weather.

63. *Galvanized Material.*—Galvanized material shall be in accordance with the Specification for Galvanizing Iron and Steel.

Bolt holes in galvanized material shall be made before galvanizing. Sherardizing for small parts is permissible.

Foundations

64. The foundations for steel poles and towers shall be designed to prevent overturning.

The weight of concrete shall be assumed as 140 lb. per cubic foot. In good ground, the weight of "earth" (calculated at 30° from the vertical) shall be assumed as 100 lb. per cubic foot. In swampy ground, special measures shall be taken to prevent uplift or depression.

Concrete for foundation shall be well worked, very wet, and shall not be leaner than one part Portland cement, three parts clean, sharp sand, and six parts of broken stone, or one part Portland cement to six parts of good gravel, free from loam or clay.

65. The top of the concrete foundation, or casing, shall be not less than six (6) in. above the surface of the ground, nor less than one (1) ft. above high water, except that no foundation need be higher than

the base of the railroad company's rail, or the top of the traveled roadway.

66. When located in swampy ground, wooden crossing and next adjoining poles shall be set in barrels of broken stone or gravel, or in broken stone or timber footings.

67. When located in the sides of banks, or when subject to washouts, foundations shall be given additional depth, or be protected by cribbing or riprap.

68. All foundations and pole settings shall be tamped in six (6) in. layers, while backfilling. It is desirable in backfilling that the earth be suitably moistened.

Working Unit Stresses

Obtained by dividing the ultimate breaking strength by the factors of safety given in paragraph 37.

69. *Structural Steel:*

	Lb. per sq. in.
Tension (net section)	18,000
Shear	14,000
Compression	$18{,}000 - 60\dfrac{l}{r}$

70. *Rivets, Pins:*

	Lb. per sq. in
Shear	10,000
Bearing	20,000
Bending	20,000

71. *Bolts:*

	Lb. per sq. in.
Shear	8,500
Bearing	17,000
Bending	17,000

72. *Wires and Cables:*

	Lb. per sq. in
Copper, hard-drawn, solid, B. & S. gage 0000, 000, 00	25,000
Copper, hard-drawn, solid, B. & S. gage 0	27,500
Copper, hard-drawn, solid, B. & S. gage No. 1	28,500
Copper, hard-drawn, solid, B. & S. gage Nos. 2, 4, 6.	30,000
Copper, soft-drawn, solid	17,000
Copper, hard-drawn, stranded	30,000
Aluminum, hard-drawn, stranded, B. & S. gage under 0000	12,000
Aluminum, hard-drawn, stranded, B. & S. gage No. 0000 and over	11,500

73. *Untreated Timber:*

	Lb. per sq. in.	$1 - \dfrac{L}{60D}$
Eastern white cedar	600	600
Chestnut	850	850
Washington cedar	850	850
Idaho cedar	850	850
Port Orford cedar	1150	1150
Long-leaf yellow pine	1000	1000
Short-leaf yellow pine	800	800
Douglas fir	900	900
White oak	950	950
Red cedar	700	700
Bald cypress (heartwood)	800	800
Redwood	650	650
Catalpa	500	500
Juniper	550	550

L = Length in inches.

D = Least side, or diameter, in inches.

GENERAL SPECIFICATIONS FOR ELECTRIC LIGHT AND POWER LINES

BY R. D. COOMBS

CLEARANCES

1. *Conductors.*—The clear headroom above a highway, under the most unfavorable condition of temperature and loading, shall be not less than 20 ft.

2. The vertical overhead clearance from any telephone or similar wire on the power line, shall be not less than:

Line voltage	Clearance
Not exceeding 6,600 volts	2 ft.
Exceeding 6,600 but not exceeding 22,000	4 ft.
Exceeding 22,000 but not exceeding 45,000	5 ft.
Exceeding 45,000 but not exceeding 66,000	6 ft.
Exceeding 66,000 but not exceeding 88,000	7 ft.
Exceeding 88,000 but not exceeding 110,000	8 ft.
Exceeding 110,000	10 ft.

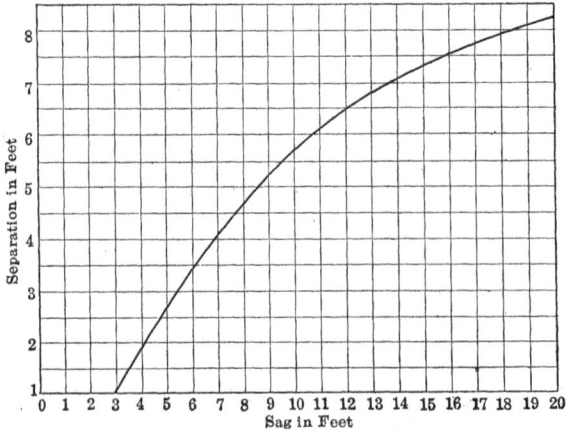

FIG. 1.—Conductor separations.

3. The separation of alternating-current conductors on pin-type insulators in a horizontal plane shall be in general not less than that required for the sag in question in Fig. 1 nor less than:

258

Line voltage	Spacing
Not exceeding 6,600 volts..........................	12 in.
Exceeding 6,600 but not exceeding 13,000...........	18 in.
Exceeding 13,000 but not exceeding 22,000...........	24 in.
Exceeding 22,000 but not exceeding 33,000...........	30 in.
Exceeding 33,000 but not exceeding 45,000...........	40 in.
Exceeding 45,000 but not exceeding 66,000...........	50 in.
Exceeding 66,000 but not exceeding 88,000...........	60 in.

4. The separation of conductors supported by suspension type insulators, in a horizontal plane, shall be that required in paragraph 3, plus one and one-quarter times the length of the suspension string.

5. The clearance between a conductor and any part of the structure shall be not less than:

Line voltage	Clearance
Not exceeding 6,600 volts..........................	9 in.
Exceeding 6,600 but not exceeding 13,000...........	12 in.
Exceeding 13,000 but not exceeding 22,000...........	15 in.
Exceeding 22,000 but not exceeding 33,000...........	18 in.
Exceeding 33,000 but not exceeding 45,000...........	21 in.
Exceeding 45,000 but not exceeding 66,000...........	24 in.
Exceeding 66,000 but not exceeding 88,000...........	27 in.
Exceeding 88,000 but not exceeding 110,000..........	29 in.
Exceeding 110,000.................................	30 in.

NOTE.—This requirement does not apply to the distance between the crossarms for an insulator of the through-pin type.

6. The side clearance between a conductor supported by suspension insulators and the supporting structure when the insulator string is deflected 45°, shall be not less than that specified in paragraph 5.

7. *Ground Wire.*—The longitudinal overhead ground wire or wires shall be in general not more than 45° from the vertical through the adjoining conductor, and with a separation of not less than that required by the table in paragraph 3.

LOADS AND FACTORS OF SAFETY

8. *Ice and Wind Loads.*—The conductors shall be considered as uniformly loaded throughout their length, with a load equal to the resultant of the dead load plus the weight of a layer of ice ½ in. in thickness and a wind pressure of 8 lb. per square foot on the ice-covered diameter, at a temperature of 0°F.

9. The weight of ice shall be assumed as 57 lb. per cubic foot (0.033 lb. per cubic inch).

10. The wind pressure on poles or towers shall be assumed at 8 lb. per square foot, on the projected area of solid or closed poles, and on one and one-half times the projected area of latticed poles, and on twice the projected area of wide base structures.

In regions in which there is no sleet, the ice load may be omitted and the wind pressure increased to 15 lb. per square foot.

11. *Conductors and Ground Wires.*—For spans exceeding 150 ft. and for lines not on streets, the conductors and overhead ground wires shall be designed to withstand the above ice and wind loads, without exceeding the elastic limit of the material.

12. For spans not exceeding 150 ft. and for all lines on streets, the above loading may be reduced 25 per cent.

13. *Ground-wire Connections.*—The ground wire connections shall be designed to withstand the maximum stress in the ground wire, without exceeding the elastic limit of the material.

14. *Supports.*—For spans exceeding 150 ft. and for lines not on streets the poles or towers shall be designed to withstand the combined stresses from their own weight, the wind pressure on the structure and the above wire loading on the adjoining spans combined with the effect of one broken conductor, with a factor of safety of 2.0.

15. For spans not exceeding 150 ft. and all lines on streets, the supporting structures shall be designed to withstand the above loading.

NOTE.—Guys may be used to obtain the strength required by paragraphs 14 and 15.

16. *Insulators and Pins.*—Insulators and pins at corners or bends in the line shall be designed to withstand the transverse loads resulting from the above ice and wind loads on the conductors, combined with the horizontal component due to the tension in the wires and the angle in the line, with a factor of safety of one and one-half (1.5).

NOTE—Double arms may be used to obtain the requisite strength.

17. *Suspension Type Insulators.*—Suspension type insulators and their connections shall be designed to withstand the maximum tension in the conductors, with a factor of safety of one and one-half (1.5) when used in the suspension position and 2.0 when used in the strain position.

18. *Guy Insulators.*—Strain insulators for guys shall be designed to withstand the maximum stress in the guy, with a factor of safety of two (2).

19. *Guys.*—Guys shall be designed to withstand their maximum stress with a factor of safety of two (2).

20. *Guy Anchorages.*—Guy anchorages shall be designed to withstand the maximum stress in the guys with a factor of safety of one and one-half (1.5).

21. *Foundations.*—The foundations of unguyed poles and towers shall be designed to resist overturning, with a factor of safety of two (2).

22. *Temperature.*—In the computation of stresses and clearances and in erection, provision shall be made for a variation in temperature from − 20°F. to 120°F. A suitable modification in the temperature require-

ments may be made for regions in which the above limits would not fairly represent the extreme range of temperature.

23. *Guys or Special Supports.*—Guys, or supporting structures of greater strength than required to withstand the preceding loads, shall be installed approximately as follows:

Wooden poles—

Side guy—all bends or corners.

Head guy—steep hills.

Head guy—unusually long spans.

Head and side guy—light lines every 1500 ft.

Head and side guy—heavy lines every 1000 ft.

Flexible structures—

Side guy—corners over 5°.

Head guy—steep hills.

Head guy—unusually long spans.

Head guy—or special structure, every 2000 ft.

Steel or concrete poles—

Side guy—sharp corners.

Head guy—steep hills.

Head guy—unusually long spans.

Head guy—light lines every 3000 ft.

Head guy—heavy lines every 2000 ft.

Rigid towers—

Special structure—sharp corners.

Special structure—every mile.

MATERIAL

24. *Overhead Ground Wire.*—The material of ground wires shall be copper, copper-covered steel, galvanized iron, galvanized steel or an approved alloy; sizes over No. 4 B. & S. gage shall be stranded.

NOTE.—The use of galvanized steel ground wire is not recommended except in sizes ⅜ in. or more in diameter.

25. The attachment of the ground wire to the structure shall be by means of a smooth grip with well-rounded ends, and a contact length of not less than three inches (3 in.).

26. *Conductors.*—Conductors shall be of copper, aluminum or other approved material.

27. *Insulators.*—Insulators shall be of porcelain, glass or other approved material.

28. *Wooden Pins.*—Wooden pins shall be sound, straight grained yellow or black locust or other approved species, free from knots over ⅛ in. in diameter, except on the shoulder or lower half of the shank, and free from checks, sapwood and worm holes.

29. *Metal Pins.*—For voltages over 13,000, insulator pins shall be of

steel, wrought iron, malleable iron, an approved metal or alloy or a combination of steel with wood, metal or porcelain.

NOTE.—Wood pins may be used for higher voltages in regions having favorable climatic conditions.

30. *Wooden Crossarms.*—Wooden crossarms shall be of seasoned timber, reasonably straight grained, out of wind, free from large, loose or unsound knots, wane, large pitch pockets, pitch pockets which enter the pin or bolt holes, through shakes, shakes or checks over 3 in. long and rot or worm holes.

31. *Wooden Poles.*—Wooden poles shall be of approved species of timber, peeled, with trimmed knots, reasonably straight, well proportioned from butt to tip, with squared ends, free from defects which would materially decrease their strength or durability and of not less than 7-in. minimum diameter at the top.

32. *Guys.*—Guys shall be stranded, galvanized steel or copper covered cable, not less than $\frac{5}{16}$ in. in diameter.

33. *Pole Steps.*—Pole steps shall be of forged or rolled iron or steel, in accordance with the Manufacturers' Standard Specification.

34. *Steel.*—Structural steel work shall be of open-hearth steel, in accordance with the Manufacturers' Standard Specification.

STRUCTURAL DESIGN

35. *Frame.*—The form of the frame shall be such that the stresses may be computed with reasonable accuracy.

36. The sections used shall permit inspection, cleaning and painting and shall be free from pockets in which water or dirt can collect.

37. The length of a main compression member shall not exceed 125 times its least radius of gyration. The length of a secondary compression member shall not exceed 180 times its least radius of gyration.

38. *Minimum Sections and Connections.*—The minimum thickness of metal shall be one-quarter inch ($\frac{1}{4}$ in.) for main members and three-sixteenth inch ($\frac{3}{16}$ in.) for secondary members.

In wide-base structures the minimum angle shall be not less than $1\frac{1}{2} \times 1\frac{1}{2} \times \frac{3}{16}$ in., and the minimum main bracing connections shall be two bolts.

39. *Rivets and Bolts.*—The minimum diameter of rivets and bolts shall be one-half inch ($\frac{1}{2}$ in.).

40. The diameter of a rivet or bolt hole shall not exceed the diameter of the rivet or bolt by more than one-sixteenth of an inch ($\frac{1}{16}$ in.).

41. The distance center to center of rivet or bolt holes shall be not less than:

Diameter of bolt or rivet	Spacing
$\frac{1}{2}$ in.	$1\frac{3}{8}$ in.
$\frac{5}{8}$ in.	$1\frac{3}{4}$ in.
$\frac{3}{4}$ in.	$2\frac{1}{8}$ in.
$\frac{7}{8}$ in.	$2\frac{1}{2}$ in.

42. *End and Edge Distances.*—The distance from the center of a bolt or rivet hole to a rolled edge or to a sheared end shall be not less than:

Diameter of bolt or rivet	Edge distance	End distance
½ in.	⅝ in.	¾ in.
⅝ in.	¾ in.	⅞ in.
¾ in.	1 in.	1⅛ in.
⅞ in.	1⅛ in.	1¼ in.

43. *Rods.*—The minimum diameter of rod bracing shall be one-half inch (½ in.).

44. Main diagonal rod bracing shall be provided with adjustable end connections having right and left threads.

PROTECTIVE COATINGS

45. All structural steel shall be thoroughly cleaned at the shop and galvanized or given one coat of approved paint.

NOTE.—In view of the thin sections used in the class of work covered by these specifications, the cleaning and painting or galvanizing required herein will be rigidly enforced. The make and brand of paint, or the mixture to be used, for both shop and field coats, shall be given in a written notification, a copy of which may be furnished the paint manufacturer.

46. Metal pins shall be galvanized or otherwise protected from corrosion.

47. *Hardware.*—All bolts, braces, lag screws, washers, etc., used on wooden or reinforced-concrete poles, shall be galvanized or sherardized in accordance with the Standard Specifications for galvanizing.

48. Guys shall be galvanized or copper-covered.

49. If required, wooden crossarms shall be treated with an approved preservative or given two coats of approved paint.

50. Wooden pole tops, crossarm gains and bolt holes shall be treated with paint or preservative.

NOTE.—The application at the ground line of at least a double-brush treatment with preservative is recommended.

51. *Painted Material.*—All contact surfaces shall be given one coat of paint before assembling.

52. All painted structural steel shall be given one field coat of an approved paint.

53. The surface of the metal shall be thoroughly cleaned of all dirt, grease, scale, etc., before painting, and no painting shall be done in freezing or rainy weather.

54. *Galvanized Material.*—Galvanized material shall be in accordance with the Standard Specification for galvanizing.

55. The spelter material shall be Prime Western for structural steel and Grade A for wire, or equal.

Foundations

56. In swampy or otherwise uncertain ground, the line supports shall be provided with broken stone, gravel, concrete or timber footings, and when located in the sides of banks or subject to washouts, shall be given additional depth or protected by cribbing, riprap, etc.

57. When possible, the top of a concrete foundation or casing shall extend not less than six inches (6 in.) above the ground nor less than one foot (1 ft.) above high water.

58. The top of the foundation shall slope down toward the sides, and be built up in the corner of angles, etc., to provide efficient drainage.

59. The thickness of the concrete casing around the butt of a steel pole shall be not less than three inches (3 in.).

60. *Wooden Pole Settings.*—Poles shall be set in the ground to depths not less than those specified in the following table:

Depth of Setting

Total length of pole (ft.)	Straight lines (ft.)	Curves, corners and points of extra strain (ft.)
30	5.0	6.0
35	5.5	6.0
40	6.0	6.5
45	6.5	7.0
50	6.5	7.0
55	7.0	7.5
60	7.0	7.5
65	7.5	8.0
70	7.5	8.0
75	8.0	8.5
80	8.0	8.5

61. *Steel and Concrete Pole Settings.*—The penetration below ground of a steel or concrete pole shall be not less than given in paragraph 60, except that concrete poles or steel poles of greater butt diameter than twenty inches (20 in.) incased in concrete, may have nine inches (9 in.) less penetration than otherwise provided.

62. *Towers or Wide Base Structures.*—Wide base towers or structures not provided with a web system below ground, shall be secured in a foundation designed to resist lateral movement at the ground line, either by the use of sufficient superficial area, concrete or cribbing.

63. The penetration shall be not less than six feet (6 ft.).

64. The anchorage plate shall be designed to withstand the maximum stress in the main leg, with a factor of safety of two (2), and in general shall be not less than four hundred square inches (400 sq. in.) in area.

65. *Excavation.*—The bottom of the excavation shall be compacted and if required shall be covered with a rammed layer of broken stone and sand, or gravel, or be covered with concrete.

Concrete

66. *Cement.*—The cement shall be Portland, and shall meet the requirements of the Standard Specifications.

67. *Aggregates.*—Aggregates shall consist of sand, gravel, broken stone or other approved material, graded from fine to coarse, free from vegetable matter and soft particles and reasonably clean.

68. *Water.*—Water used in mixing concrete shall be free from oil, acid and injurious amounts of alkalies or vegetable matter.

69. *Proportions.*—For plain concrete or mass foundations, not less than one part cement to a total or nine (9) parts of fine and coarse aggregates, measured separately, shall be used.

70. For reinforced concrete not less than one (1) part of cement to a total of six (6) parts of fine and coarse aggregates, measured separately, shall be used.

71. Such relative amounts of fine and coarse aggregates shall be used as will produce a dense uniform concrete.

72. *Mixing.*—The materials shall be well mixed, using sufficient water to form a mixture wet enough to flow in the forms and about the reinforced or incased metal, but which will not permit the separation of the coarser aggregates from the mortar.

73. *Workmanship.*—Proper precautions shall be taken to prevent the freezing of concrete.

74. Mortar or concrete which has partially set shall not be remixed and used.

75. Exposed surfaces of the concrete shall be rubbed smooth without plastering.

76. Exposed edges or corners shall be rounded or beveled.

77. *Concrete Poles.*—Reinforced-concrete poles shall be made strictly in accordance with the best practice in workmanship, using approved aggregates and producing a concrete of great density.

78. No part of the reinforcement shall be less than one inch (1 in.) from the surface.

79. No cast metal shall be used in the skeleton reinforcement.

80. The reinforcing rods shall be capable of being bent cold 180° around a circle of four diameters.

81. The entire surface of the concrete shall be rubbed to a smooth finish without plastering.

82. Poles shall be straight, of the required dimensions, and provided with the necessary holes for crossarms, brace and guy bolts and sockets for pole steps.

Tests

83. *Insulators.*—Each insulator shall be so designed that, with excessive potential, failure will first occur by flash-over and not by puncture.

84. Previous to the electrical tests the separate parts of an insulator shall be subject to inspection for mechanical defects in material or workmanship. No part shall contain soft porcelain, crazing, serious deformations or cracks in the grooves or in the unglazed portions that would materially decrease the value of the insulator.

85. The assembled insulators shall withstand a voltage of 5000 volts less than specified in paragraphs 90 and 91, for five consecutive minutes, without injury, abnormal static strain, noise, or flash-over.

86. Each separate part of a built-up insulator, and each assembled and cemented insulator shall be subjected to an approved factory test.

87. The wet flash-over test shall be made under a precipitation of water of one-fifth of an inch per minute, at an angle of 45° to the axis of the insulator.

88. Test voltages above 35,000 volts shall be determined by the A.I.E.E. Standard Spark-gap Method.

89. Test voltages below 35,000 volts shall be determined by transformer ratio.

90. *Pin Insulators.*—The flash-over design test voltage shall be not less than:

Line voltage	Dry flash-over	Wet flash-over
Less than 11,000	. . . . . .	twice line voltage
11,000	60,000	30,000
22,000	90,000	50,000
33,000	100,000	60,000
45,000	125,000	90,000
50,000	150,000	100,000
60–70,000	180,000	120,000
80,000	240,000	160,000

91. *Suspension Type Insulators.*—The flash-over design test voltage of the string of suspension units shall be not less than:

Line voltage	Dry flash-over	Wet flash-over
11,000	80,000	50,000
22,000	160,000	90,000
33,000	160,000	90,000
45,000	220,000	130,000
66,000	270,000	175,000
88,000	310,000	220,000
110,000	340,000	265,000
125,000	460,000	300,000
140,000	470,000	335,000

92. When insulators of the suspension type are placed in the strain position, one additional insulator unit shall be used in series.

93. For line voltages not exceeding 9000 volts, strain insulators for guys·shall have a wet flash-over of not less than four times the maximum line voltage.

94. *Reinforced-concrete Poles.*—If required, the strength of concrete poles shall be determined by testing one pole from each lot of......

95. The expense of testing the poles which withstand the specified conditions of loading shall be borne by the purchaser and the expense of testing those which do not meet the requirements shall be borne by the contractor.

96. Poles shall be tested by applying a horizontal pull at the center of gravity of the wires, and continually increasing the stress until it is equivalent to the specified loading.

97. Test poles shall be set in a firm foundation.

98. The test load shall be measured by a dynamometer, or scale.

INDEX